TOPICS IN COLLEGE MATHEMATICS

THIRD EDITION

David C. Mello, Ph.D.
JOHNSON & WALES UNIVERSITY

This book was set in Palatino by Hermitage Publishing Services and printed and bound by
Von Hoffmann Press, Inc.

This book is printed on acid-free paper. ♾

ISBN 0471-22975-X

Printed in the United States of America

10 9 8 7 6 5 4 3 2 1

CONTENTS

To the Instructor

This book grew out of a set of lecture notes for a course entitled "A Survey of College Mathematics" which is offered at Johnson & Wales University. Its goal is to present the basic elements of college-level mathematics in an easy to read, but rigorous style.

In keeping with the original philosophy of the first edition, the new edition has been written with beginners in mind. All of the basic requirements for each topic are carefully developed in each lesson, and each example is solved in great detail.

In addition, lofty assumptions have not been made about the level of mathematical sophistication possessed by the student; in fact, it is only assumed that your students have a working knowledge of elementary arithmetic, and possess a strong desire to learn the subject material.

Unlike most elementary texts of this type, proofs have been provided to most of the theorems presented in the text; these proofs have been provided to encourage critical thinking, and for several other important reasons.

As a fellow instructor, I have experienced that many beginning students suffer from the false impression that the study of mathematics merely consists of an arbitrary collection of rules and procedures that must be memorized in order to solve useful problems. It is felt that the inclusion of proofs will afford students with an appreciation of the fact that mathematics is not an arbitrary endeavour, but a creative science which is built upon the bedrock of logic.

If your better students choose to work through the demonstration of a given theorem, then they will gain a deeper understanding of the concepts employed, and can then draw deeper connections between existing concepts and others.

The book has been organized into "lessons" rather than chapters. This has been done to limit the size of the mathematical morsels that must be digested your students during each class, and to make it easier for you, the instructor, to budget class time. Most lessons can be covered in a 55 minute class period.

Considerably more material has been included in the book than is normally covered in an elementary course of this nature. Supplementary topics such as quadratic equations, matrices, and the mathematics of finance have been included. Such topics may be covered as time permits, and according to your interests and those of the class.

The third edition also includes three additional lessons on quadratic functions and the algebra of complex numbers, and nearly 200 new problems have been added to the original text.

Each set of exercises has been divided into two parts: Part A and Part B. Exercises falling under the former designation are largely routine in nature and can be completed by students with a minimum of effort.

Part B exercises are usually more challenging and give the student numerous opportunities to apply newly acquired concepts in unfamiliar settings. Part B exercises may be appropriate for those students who elect to take the course with an honors option.

I would like to thank Professor Joseph Alfano, Professor Richard Cooney, Professor Mark Duston, Professor Evelyn Giusti, Professor Gail St. Jacques, Dr. Joyce Oster, Professor Lucy Ligas, Professor Charles Mazmanian, Professor Thomas Pandolfini, Dr. Prem Singh, Professor Carmine Vallese, Dr. Garri Yuzevovich, and Professor Sandra Weeks, for their assistance in reviewing the original manuscript, and for their many helpful comments which arose from their use of the first edition.

Finally, I am also grateful to the many students of Johnson & Wales who took the time to share their impressions, and gave many suggestions for improving the first edition of the text.

David C. Mello
Providence, Rhode Island
June, 2002

To the Student

This book has been written with you, the student, in mind. It has been designed to help you learn the basic elements of college mathematics in an enjoyable fashion. Hopefully, it will give you some glimpse of the intrinsic beauty of what we call "mathematics" and how mathematics can be used in your chosen career.

In using the book, you should pay close attention to the definitions of new concepts, and to the various theorems; these items have been clearly designated throughout the text. A **theorem** is nothing more than a useful mathematical statement whose truth must be established or "proved."

Whenever possible, you should try to work through the proof of each theorem. If you take the time to do this, you will be rewarded with a deeper understanding of the material, and you will find it less necessary to memorize key points in preparation for your examinations.

Learning mathematics is like learning to play a new musical instrument for the first time. Listening to someone else play a beautiful melody may be enjoyable, but it doesn't help you master the instrument yourself. You have to take the time to actually "do mathematics" in order to play its music.

But exactly how do you "do mathematics?" Here are some helpful hints for learning and doing mathematics:

1. Always read the text with a pencil in hand, and take the time to work through the details of each example provided in the text. Numerous examples have been provided to help you master the course material.
2. Don't be afraid to write in the text; the book is meant to be used. Its fairly wide margins will give you some space for short notes and for "scratch work."
3. Be sure to do the homework problems that are assigned by your instructor. If you are unable to do some of the assigned work, or if you are confused about some point, then don't get discouraged, but be sure to ask your instructor for help.
4. Prepare for each class by reading the assigned lesson in the text *before class*. If you take the time to do this, you'll find that you will spend less

time taking notes, and more time really understanding and enjoying your classes.

5. Keep a complete but *concise* notebook for the course. Along with the text, this will be a valuable aid in reviewing for your examinations.

I sincerely hope that you will find the textbook to be extremely helpful and that you will enjoy "doing mathematics." Don't be afraid of mathematics; a great physicist once said that "if you give mathematics a chance, and if you learn to love mathematics, then it will love you back!"

1 INTRODUCTION TO SETS

The concept of a **set** plays an important role in modern mathematics. If we were to imagine that the whole of Mathematics were a huge house of cards, then the cards corresponding to the set idea would be at the very foundation of this fragile structure; that is how fundamental this concept really is to our ability to do meaningful mathematics. This leads us to the obvious question: "What is a set?"

Definition-1: (Set)

A **set** is just a collection of objects that are called **members** or **elements** of the set.

Mathematicians use capital letters to denote sets, and sometimes, the easiest way to **define** a set is to simply list its elements. We list the elements of any given set between two braces, and separate its elements by commas. When we list the elements of a given set, we shall agree that the **order** in which we list these elements is immaterial. We call this method the **listing method,** and we give several examples of its use below:

Example-1: Let D be the set of the first three days of the week. Use the listing method to define D.

Solution:

$$D = \{Sunday, Monday, Tuesday\}$$

Example-2: If F is the set of the first five counting numbers, then use the listing method to define F.

Solution:

$$F = \{1,2,3,4,5\}$$

Example-3: Let V be the set of vowels in the English alphabet. Use the listing method to define V.

Solution:

$$V = \{a,e,i,o,u\}$$

Example-4: Suppose that E is the set of all even counting numbers greater than 3 but less than 15. Use the listing method to define E. [Remember, an even number is one which is divisible by two.]

Solution:

$$E = \{4,6,8,10,12,14\}$$

The above examples illustrate that the elements of a set can be anything; they don't have to be numbers. However, an important set of numbers that we use on a daily basis is the set of **natural numbers** or **counting numbers:**

THE SET OF NATURAL NUMBERS

$$N = \{1,2,3,4,5,6,...\}$$

Here, we have used an **ellipsis** (i.e., three dots) after the 6 to indicate that the set does not terminate, and continues in a similar manner after the element 6. Similarly, an ellipsis can be used **between** two given elements to make the listing process easier. Take a look at the next example.

Example-5: Let B denote the set of natural numbers greater than 17 but less than 101. Use the listing method to define this set.

Solution:

$$B = \{18,19,20,21,...,100\}$$

Mathematicians are very lazy people. This means that they always use special symbols wherever possible to shorten their work. For example, they use the special symbol $\in$ to indicate membership in a set, and the symbol $\notin$ to indicate that a given element does not belong to a certain set. Thus, if A is any set, and x is some object which may or may not belong to A, we shall agree that:

MEMBERSHIP NOTATION

$x \in A$ means "x is an element of the set A"

$x \notin A$ means "x is not an element of the set A"

Example-6: Given that

$$A = \{1,2,3,...,10\}$$

We see that:

$$1 \in A, \quad 2 \in A, \ldots, 10 \in A$$

but, for example,

$$11 \notin A \text{ and } 17 \notin A$$

An alternative way of defining a set is to use what is called **set builder notation.** This method is often useful when the given elements of a set cannot be conveniently listed. The general format of this notation is as follows:

SET BUILDER NOTATION

$$A = \{x \mid x \text{ (satisfies some condition to be an element of A)}\}$$

Which is read:

A is the set of all elements x such that x is . . .
In this notation, our lazy mathematicians have used the symbol " | " as shorthand for the phrase "such that." This method of defining a set may look awkward at first, but some simple examples will help us to grasp it.

Example-7: Let F be the set of the first five natural numbers. Use set-builder notation to define F.

Solution:

$$F = \{x \mid x \in N \text{ and } x < 6\}$$

Here, F is "the set of all elements x such that x is a natural number and is less than 6."

Example-8: Use set-builder notation to define the set B in Example-5 above.

Solution:

$$B = \{x \mid x \in N \text{ and } 17 < x < 101\}$$

In this case, B is "the set of all elements x such that x is a natural number and, x is greater than 17 and less than 101."

Example-9: Using set-builder notation, define the set *E* of all even natural numbers.

Solution:

$$E = \{x \mid x \in N \text{ and } x \text{ is even}\}$$

In each of the above examples, the sets we have used have been **well defined;** that is, if we are presented with any object, it is possible for us to determine whether or not that object belongs to each given set. An example of a set which is not well-defined is: "The set of the top ten Rock guitarists in the world." It is easy to see that we could argue endlessly on whether or not any given guitarist belongs to this set. For this reason, we will only consider well defined sets in all of our future work!

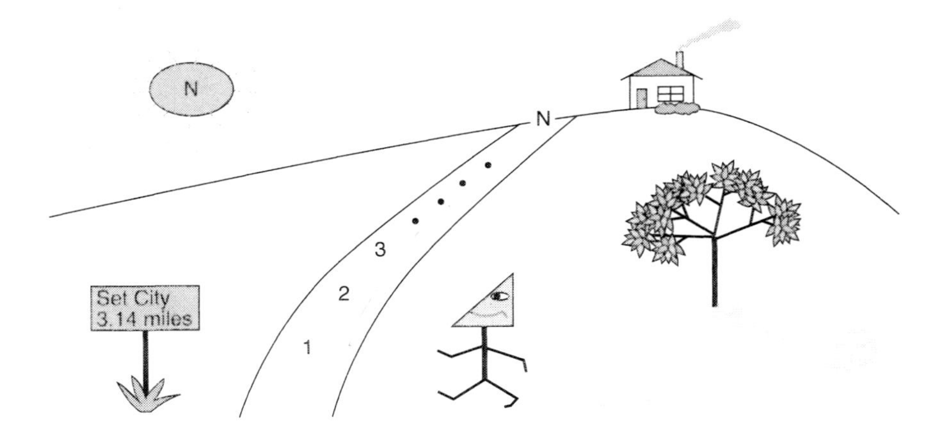

Figure 1. Triangle Man on his way to Set City

EXERCISE SET #1

PART A

In Exercises 1–7, use the listing method to define each of the following sets:

1. The set of all natural numbers less than 5.
2. The set of all college courses you are taking this term.
3. The set of all even numbers greater than 2 but less than 10.
4. The set of all natural numbers less than 20 and that are evenly divisible by 3.
5. The set of all odd natural numbers greater than 10.
6. The set of all roommates in your dormitory room.
7. The set of all prime numbers less than 30. [A prime number is a natural number which is evenly divisible *only* by itself and one. For example, 2,3,5 are the first three primes.]

In Exercises 8–13, indicate whether each of the following statements are true or false:

8. $1 \in \{1,2,3,4,5\}$
9. $2 \in \{-1,3,0\}$
10. $0 \notin N$
11. $N = \{1,2,3,4,\ldots\}$
12. $\frac{3}{2}$ is a natural number.
13. The expressions: $\{a,b,c\}$ and $\{b,a,c\}$ represent the same set.

In Exercises 14–25, use set-builder notation to define each of the following sets:

14. The set of all natural numbers less than 10.
15. The set of all odd numbers greater than 3 but less than 12.
16. The set of all natural numbers that are evenly divisible by 4.
17. The set of all odd natural numbers greater than 10.
18. The set of all states in the United States.
19. The set of all even natural numbers less than 9.
20. The set of all odd natural numbers less than 7.
21. The set of all prime numbers less than 10.
22. The set of all natural numbers divisible by 3.
23. The set of all natural numbers greater than 2 but less than 13.
24. The set of all odd natural numbers greater than 11.
25. The set of all natural numbers divisible by 5.

2 More About Sets

In this lesson, we shall further explore the basic properties of sets. You have probably already noticed that there are two basic types of sets: **finite sets** and **infinite sets.**

Definition-1: (Finite and Infinite Sets)

A set A is said to be a **finite set** if and only if when attempting to count the number of elements in A the counting process eventually *terminates*. If this is not the case, we say that A is an **infinite set.**

A different way of saying the same thing is that a given set A is finite if the number of elements in A is a natural number; otherwise, we say that the given set is infinite. It is clear that the set of natural numbers is an example of an infinite set while the set of letters in the English alphabet is a finite set.

The number of elements in a given finite set is called its **cardinality** or **cardinal number.** If two sets have the same number of elements, i.e., if they have the same cardinality, then we say that they are **equivalent** sets. We summarize these definitions below:

Definition-2: (Cardinality of a Set)

The **cardinality** of a finite set A, denoted by $\#(A)$, is the number of elements which A contains.

Definition-3: (Equivalent Sets)

The finite sets A and B are said to be **equivalent** if and only if they contain the *same number* of elements; that is, if:

$$\#(A) = \#(B)$$

Let's look at an example in order to clarify these concepts.

Example-1: Given the sets:

$$A = \{1,2,3,4,5\} \text{ and } B = \{a,e,i,o,u\}$$

we see that both A and B are finite sets. Furthermore, we have:

$$\#(A) = 5 \text{ and } \#(B) = 5$$

so we conclude that A and B are equivalent.

We say that a set A is **equal** to a set B if both sets have the same elements. This means that A and B are equal if and only if *every* element of A is also and element of B, and conversely, *every* element of B is also an element of A.

Definition-4: (Equality of Sets)

The sets A and B are said to be **equal,** and we write:

$$A = B \quad \text{(which is read "A equals B")}$$

if and only if they both have *exactly* the same elements.

After some thought, the reader should realize that two sets can be equivalent but not equal to each other; be careful not to confuse these two concepts. On the other hand, if two sets are equal to each other, then they must contain the same number of elements, and so they are equivalent. Let's look at some examples.

Example-2: Consider the sets:

$$A = \{2,3,4\} \text{ and } B = \{4,2,3\}$$

We see that A and B contain exactly the *same* elements so that these sets are equal, and it is permissible to write $A = B$.
Observe that the *order* in which the elements appear in each set doesn't matter, only the contents (elements) of each set are of importance. Incidentally, since:

$$\#(A) = \#(B) = 3$$

we can conclude that A and B are equivalent as well.

Example-3: Given the sets:

$$E = \{2,4,6,8,10,...\} \text{ and } N = \{1,2,3,4,5,...\}$$

The set E is just the set of even natural numbers while N is the entire set of natural numbers. It is clear that every element of E is also an element of N but the converse is not true; that is, there are some elements in N that are *not* elements of E. For example,

$$1 \in N \quad \text{but} \quad 1 \notin E$$

$$3 \in N \quad \text{but} \quad 3 \notin E$$

and so on. We conclude that these sets are not equal, and write $E \neq N$.

It may seem surprising at first, but there are some sets that don't contain any elements at all! We call such sets **empty sets** or **null sets** and denote all such sets by the same symbol: $\emptyset$. Consider the following examples:

Example-4: Consider the set S where

$$S = \{x \mid x \in N \text{ and } 2x = 5\}$$

We might describe S as the set of all natural numbers x which satisfy the equation: $2x = 5$. It is clear that $x = 5/2$ is the only number which satisfies this equation since:

$$2\left(\frac{5}{2}\right) = 5$$

Since 5/2 is not a natural number, i.e., $5/2 \notin N$, we conclude that S can't contain any elements, i.e., S is an empty set, and we write $S = \emptyset$.

Example-5: Given the set:

$$T = \{x \mid x \text{ is a triangle and } x \text{ has four sides}\}$$

From elementary geometry, we know that a triangle cannot have four sides by its very nature. Thus, T is an empty set and it is permissible to write $T = \emptyset$ as well.

From the above examples, you should observe that there is more than one empty set; in fact, there are infinitely many such sets. Furthermore, since:

$$\#(S) = \#(T) = 0$$

then we can conclude:

$$\#(\emptyset) = 0 \tag{1}$$

for any empty set, and all empty sets are equivalent.

Definition-5: (Empty Set)

A set is said to be an **empty set** (or **null set**) if it doesn't contain any elements. We agree to denote any empty set by the symbol: $\emptyset$ or $\{\ \}$.

The student should be careful not to use the symbol $\{\emptyset\}$ to represent an empty set. This set is not an empty set, but the set which contains the empty set as its only *element*. Also, notice that:

$$\#(\{\emptyset\}) = 1$$

so this set cannot be empty.

EXERCISE SET #2

PART A

In Exercises 1–7, use the following sets:

$$A = \{a,b,c,d\}$$
$$B = \{1,2,3,4\}$$
$$C = \{x \mid x \in N \text{ and } x \leq 0\}$$
$$D = \{x \mid x \in N \text{ and } x < 6\}$$
$$E = \{x \mid x \text{ is a triangle}\}$$

to determine whether each of the following statements is true or false.

1. A and B are equivalent sets.
2. $A = B$
3. C is an empty set.
4. $\#(A) = \#(B) = 4$
5. $\#(E) = 3$
6. E is an infinite set.
7. $\#(C) = 0$

In Exercises 8–14, determine whether each set is finite or infinite:

8. $A = \{x \mid x \in N \text{ and } x > 6\}$
9. $B = \{x \mid x \in N$ and x is evenly divisible by 3.$\}$
10. $C = \{x \mid x$ is a word in Shakespeare's play Hamlet.$\}$
11. $D = \{x \mid x$ is a possible move in a Chess game.$\}$
12. $E = \{2,4,6,8,10,12,...\}$
13. $F = \{x \mid x$ is a solution of the equation $2x = 6\}$
14. $G = \{x \mid x$ is a solution of the inequality $2x > 4\}$

In Exercises 15–18, determine whether the following pairs of sets are equal, equivalent, both, or neither:

15. $A = \{a,b,c\}$, $B = \{2,1,3\}$
16. $A = \{3,b,f\}$, $B = \{b,3,f\}$
17. $A = \{x \mid x \in N \text{ and } x \leq 5\}$ and $B = \{2,1,3,5,4\}$
18. $A = \{x \mid x \in N \text{ and } x < 5\}$, $B = \{4,1,3,2\}$

In Exercises 19–22, use the sets:

$$A = \{a,b,c,d\}$$
$$B = \{1,0,3,5\}$$
$$C = \{x \mid x \in N \text{ and } x \leq 1\}$$
$$D = \{x \mid x \in N \text{ and } 3 < x < 6\}$$

to calculate the following:

19. $\#(A)$
20. $\#(B)$
21. $\#(C)$
22. $\#(D)$

3 SUBSETS

Besides the possibility of two sets being equal or having the same number of elements, sets can be related in several other ways. If this were not true, the theory of sets would neither be interesting nor useful. For example, consider the following two sets:

$$A = \{2,3,4\} \text{ and } B = \{5,1,2,4,3\}$$

We notice that *every* element of A is also an element of B. When a situation like this arises, we say that "A is a **subset** of B" and write:

$$A \subseteq B \quad \text{(which is read "}A\text{ is a subset of }B\text{")}$$

Definition-1: (Subset)

Given the sets A and B, if every element of A is also an element of B, then we say that A is a **subset** of B (or B **contains** A) and write $A \subseteq B$. Stated differently,

SUBSET

$A \subseteq B$ if and only if $x \in A$ implies $x \in B$ (for all elements $x \in A$)

Starting with this simple definition, we can make several interesting observations. First, notice that every set must be a subset of itself; that is:

EVERY SET IS A SUBSET OF ITSELF

$$A \subseteq A \qquad \text{for } \textit{every} \text{ set } A \tag{1}$$

Secondly, by default, the empty set must be a subset of every set. That is,

THE NULL SET IS A SUBSET OF EVERY SET

$$\emptyset \subseteq A \text{ for } \textit{every} \text{ set } A \tag{2}$$

If (2) were not true, this would mean that there is some element in $\emptyset$ which is not an element of A; but this is impossible since the null set doesn't contain any elements by its very definition!

Finally, if we are given any two sets A and B, recall that these two sets are equal if and only if every element of A is also an element of B (so $A \subseteq B$) and every element of B is also an element of A (so $B \subseteq A$). This provides us with another method for checking the equality of any two given sets:

ALTERNATIVE TEST FOR EQUALITY OF SETS

$$A = B \quad \text{if and only if} \quad A \subseteq B \text{ and } B \subseteq A \tag{3}$$

Suppose we are given two sets A and B where $A \subseteq B$, but $A \neq B$. Since $A \neq B$, we are forced to conclude that there must be *at least one* element of B which is *not* an element of A. Mathematicians describe this state of affairs by saying that "A is a **proper subset** of B", or "B **properly contains** A."

Definition-2: (Proper Subset)

Given two sets A and B we say that "A is a **proper subset** of B", (or "B **properly contains** A") and we write:

$$A \subset B$$

if every element of A is also an element of B, but $A \neq B$.

Alternatively, we have:

PROPER SUBSET

$$A \subset B \text{ if and only if } A \subseteq B \text{ but } A \neq B \tag{4}$$

Example-1: Given the sets:

$$A = \{a,1,d,4\} \text{ and } B = \{d,1,4,a,g,h\}$$

determine whether $A \subset B$.

Solution: Here we notice that every element of A is also an element of B (so that $A \subseteq B$) but $A \neq B$. We conclude that $A \subset B$.

Definition-3: (Power Set)

The set of all the *distinct* subsets of any given set A is called the **power set** of A and is denoted by $P(A)$.

Example-2: Given the set:

$$B = \{a,b\}$$

Find the power set of B.

Solution: We begin by listing the possible distinct subsets of B:

$$\emptyset,\ \{a\},\ \{b\},\ \{a,b\}$$

Notice we had to include both $\emptyset$, and B itself in our list. The power set of B is then the set which contains all of the above subsets:

$$P(B) = \{\ \emptyset,\ \{a\},\ \{b\},\ \{a,b\}\ \}$$

Example-3: Consider the set:

$$S = \{1,2,3\}$$

Find the power set of S.

Solution: The only distinct subsets of S are:

$$\emptyset,\ \{1\},\ \{2\},\ \{3\},\ \{1,2\},\ \{1,3\},\ \{2,3\},\ \{1,2,3\}$$

so the power set of S is:

$$P(S) = \{\ \emptyset,\ \{1\},\ \{2\},\ \{3\},\ \{1,2\},\ \{1,3\},\ \{2,3\},\ \{1,2,3\}\ \}$$

From the above examples, it should be apparent that the power set of any finite set must be a finite set itself, although for large finite sets, it might be quite tedious to list.

Also, since the only possible subset of the null set is itself, its power set contains only one element:

$$P(\emptyset) = \{\emptyset\}$$

Now, if we are given any finite set, how can we determine the number of elements in its power set? This question is answered in the following theorem:

Theorem-1: (Number of Distinct Subsets)

If A is any *finite* set, then the cardinality of the power set of A is given by:

$$\#(P(A)) = 2^{\#(A)} \quad (5)$$

Stated differently, **if a finite set A contains n elements, then it has a total of 2^n distinct** (different) **subsets.**

Example-4: Given the set:

$$S = \{a,b,c,d\}$$

(a) How many elements does $P(S)$ contain?
(b) List the elements in the power set of S.

Solution:
(a) Since the set S contains four elements, i.e.,

$$\#(S) = 4$$

then by Theorem-1, the total number of elements in the power set of S is given by:

$$\#(P(S)) = 2^{\#(S)} = 2^4 = 2 \cdot 2 \cdot 2 \cdot 2 = 16$$

(b) The elements in the power set of S are as follows:

$$\emptyset$$
$$\{a\}, \{b\}, \{c\}, \{d\}$$
$$\{a,b\}, \{a,c\}, \{a,d\}, \{b,c\}, \{b,d\}, \{c,d\}$$
$$\{a,b,c\}, \{a,b,d\}, \{a,c,d\}, \{b,c,d\}$$
$$\{a,b,c,d\}$$

Example-5: Given the sets:

$$T = \{3,8,a,4,2\}$$
$$W = \{1,2,3,4,\ldots,8,9,10\}$$

calculate the number of elements in the power set of each set.

Solution:
Since we have:

$$\#(T) = 5 \text{ and } \#(W) = 10$$

then Theorem-1 assures us that:

$$\#(P(T)) = 2^{\#(T)} = 2^5 = 2 \cdot 2 \cdot 2 \cdot 2 \cdot 2 = 32$$

and similarly,

$$\#(P(W)) = 2^{\#(W)} = 2^{10} = 1{,}024$$

This last example shows just how useful Theorem-1 can be; listing the elements in $P(T)$ would be quite tedious, while if we attempted to list the elements in $P(W)$, this might take us a very long time indeed!

This seems to be how Mathematics often works; with just some clear definitions, and careful detective work on our part, we can often draw amazing conclusions. In Mathematics, there are free lunches!

Exercise Set #3

PART A

In Exercises 1–10, determine whether each statement is true or false:

1. $\emptyset \subseteq A$ for any set A.
2. $\{a,b\} \subseteq \{b,a\}$
3. $\{a\} \subset \{a,b,c\}$
4. {Mays, Mantle} $\subseteq$ {Mantle, Maris, Mays}
5. Mozart $\subseteq$ {Mozart, Bach, Brahms}
6. $A \subset A$ for any set A.
7. If the sets A and B are equivalent, then $\#P(A) = \#P(B)$.
8. If $A \subseteq B$, then $\#P(A) \leq \#P(B)$.
9. If $P(A) = P(B)$ then $A = B$.
10. {Einstein} $\subseteq$ {Einstein}

In Exercises 11–19, use the sets:

$$A = \{2,3,4\}$$
$$B = \{2,3,4,5,6\}$$
$$C = \{2,4,3\}$$

to determine whether each of the following statements is true or false:

11. $A \subseteq B$
12. $A \subset B$
13. $\#(P(A)) = \#(P(C)) = 8$
14. $P(A) = P(C)$
15. $C \in P(C)$
16. $\emptyset \in P(A)$
17. $\emptyset \subseteq A$
18. $A = B$
19. $A = C$
20. Given the set: $S = \{x,y,z\}$, find $P(S)$.
21. Given the set: $A = \{1,b,2\}$, find $P(A)$

PART B

22. In a certain restaurant, patrons can order their hamburger plain, or choose among any of the following set of items to put on their hamburger:

    {pickles, catsup, mustard, relish, cheese, tomato}

In how many different ways can a hamburger be ordered?

23. A person can order a new sports car with any or none of the following options: CD player, automatic transmission, leather seats, large V8 engine, convertible top, air conditioning, and a turbo-charged engine. In how many different ways can the sports car be ordered?
24. If a finite set A contains n elements, then how many *proper* subsets does A have?

4 SET OPERATIONS

In elementary arithmetic, you learned how to add, subtract, multiply, and divide two numbers. These activities are examples of what mathematicians call *operations*. An **operation** is just a *rule* which assigns a number (the "answer") to any pair of numbers that are given. For example, the operation "+" assigns the number 5 to the pair of numbers 2 and 3.

In a similar way, we can define **set operations** on any pair of sets. Unlike ordinary arithmetic operations, however, set operations don't give numbers as their answers, but *sets*. We begin this lesson with the operation of the **union** of two sets:

Definition-1: (Union of Sets)

Given the sets A and B, the **union** of A and B is denoted by

$A \cup B$ (which is read "A union B")

is the set consisting of all those objects that are elements of A, or elements of B, or both A and B.

Using set builder notation, we can define the union of two sets in a very precise way:

THE UNION OF SETS

$$A \cup B = \{x \mid x \in A \text{ or } x \in B\} \tag{1}$$

As a matter of convention, we shall agree that if a element is in common to both A and B, then we will list it only once in A ∪ B. A simple example will clarify this idea.

Example-1: Given the sets:

$$A = \{a,b\}, \quad B = \{1,2,a\}, \quad C = \{c,1,d\}$$

calculate each of the following:

(a) $A \cup B$
(b) $B \cup C$
(c) $B \cup P(A)$

Solution:

(a) We have:

$$A \cup B = \{a,b\} \cup \{1,2,a\} = \{a,b,1,2\}$$

Observe that although the element a appears in both A and B, we only need to list it once in the union of these sets.
(b) Similarly,

$$B \cup C = \{1,2,a\} \cup \{c,1,d\} = \{1,2,a,c,d\}$$

(c) In order to calculate $B \cup P(A)$, we first need to determine the power set of $\dot{A}$. After a little work, we find:

$$P(A) = \{ \emptyset, \{a\}, \{b\}, \{a,b\} \}$$

so that:

$$\begin{aligned} B \cup P(A) &= \{1,2,a\} \cup \{ \emptyset, \{a\}, \{b\}, \{a,b\} \} \\ &= \{1,2,a, \emptyset, \{a\}, \{b\}, \{a,b\} \} \end{aligned}$$

Notice that a and $\{a\}$ are different objects, so both of them have been listed as elements of $B \cup P(A)$.

The next operation that we will define is the **intersection** of sets.

Definition-2: (Intersection of Sets)

Given the sets A and B, the **intersection** of A and B is denoted by

$A \cap B$ (which is read "A intersection B")

is the set consisting of all those objects that are *both* elements of A, and elements of B.

Alternatively, using set-builder notation, we can write:

THE INTERSECTION OF SETS

$$A \cap B = \{x \mid x \in A \text{ and } x \in B\} \qquad (2)$$

Example-2: Given the sets:

$$A = \{a,b,e\}$$
$$B = \{1,a,b,c\}$$
$$C = \{d,e\}$$

calculate each of the following:

(a) $A \cap B$
(b) $A \cap C$
(c) $B \cap C$

Solution:
(a) We have:

$$A \cap B = \{a,b,e\} \cap \{1,a,b,c\} = \{a,b\}$$

(b) Similarly, we find:

$$A \cap C = \{a,b,e\} \cap \{d,e\} = \{e\}$$

(c) Here we get:

$$B \cap C = \{1,a,b,c\} \cap \{d\} = \emptyset$$

since B and C don't have any elements in common; that is, the set of all objects common to both B and C is empty.

From part (c) of the previous example, it is apparent that two given sets may not have any elements in common at all. When this is the case, mathematicians say that the sets are **disjoint.**

Definition-3: (Disjoint Sets)

Any two sets A and B are said to be **disjoint** if and only if:

$$A \cap B = \emptyset \qquad (3)$$

In a manner of speaking, we can even perform a kind of "subtraction" with any two sets. This is the notion of the **difference** between the sets A and B.

Definition-4: (Difference of Sets)

For any sets A and B, the **difference** of A and B, denoted by:

$A - B$ (which is read "A minus B")

is the set which consists of all elements of A that are **not** elements of B.

Using set-builder notation we can write:

DIFFERENCE OF SETS

$$A - B = \{x \mid x \in A \text{ and } x \notin B\} \quad (4)$$

Example-3: Given the sets:

$$A = \{1,2,3,4\}, \text{ and } B = \{2,4,6\}$$

calculate each of the following:

(a) $A - B$
(b) $B - A$
(c) $A - A$

Solution:
(a) We have:

$$A - B = \{1,3\}$$

since these are the only elements of the set A that belong to A but not to B.
(b) Similarly, we find:

$$B - A = \{6\}$$

(c) Finally, using (4), when $A = B$, we get:

$$A - A = \{x \mid x \in A \text{ and } x \notin A\}$$

Since we cannot have both $x \in A$ and $x \notin A$, we are forced to conclude that:

$$A - A = \emptyset$$

and this statement holds for *any* set A.

In our work with sets, you have probably noticed that all of the sets we discuss in any given problem are usually subsets of some larger, fixed set. We call this set the **universe of discourse,** or simply, the **universal set** for the given problem, and we agree to denote it by U.

Stated differently, a universal set is a set that contains all of the possible elements we need to handle any specific discussion involving sets. Its meaning will become clearer in the discussion that follows.

Definition-5: (Complement of a Set)

For any set A, the **complement** of A, denoted by A', is the set which consists of all of the elements of the universal set U that are *not* elements of A.

In other words, we have:

COMPLEMENT OF A SET

$$A' = U - A = \{x \mid x \in U \text{ and } x \notin A\} \tag{5}$$

Example-4: Given the sets:

$$U = \{1,2,3,4,\ldots, 10,11,12\}$$
$$A = \{1,2,8\},\ B = \{1,3,9\}$$

calculate each of the following:
(a) A'
(b) B'
(c) $(A - B)'$
(d) $(A \cap B)'$

Solution:
(a) By the definition of the complement of a set, we have:

$$\begin{aligned} A' &= \{x \mid x \in U \text{ and } x \notin A\} \\ &= \{3,4,5,6,7,9,10,11,12\} \end{aligned}$$

(b) Similarly,

$$\begin{aligned} B' &= \{x \mid x \in U \text{ and } x \notin B\} \\ &= \{2,4,5,6,7,8,10,11,12\} \end{aligned}$$

(c) In order to calculate $(A - B)'$, we must calculate what's inside the parentheses first, and then take the complement of that result. First, we get:

$$\begin{aligned} A - B &= \{1,2,8\} - \{1,3,9\} \\ &= \{2,8\} \end{aligned}$$

so that:

$$\begin{aligned} (A - B)' &= \{x \mid x \in U \text{ and } x \notin (A - B)\} \\ &= \{1,3,4,5,6,7,9,10,11,12\} \end{aligned}$$

(d) Once again, working inside the parentheses first, we obtain:

$$\begin{aligned} A \cap B &= \{1,2,8\} \cap \{1,3,9\} \\ &= \{1\} \end{aligned}$$

Consequently,

$$\begin{aligned} (A \cap B)' &= \{x \mid x \in U \text{ and } x \notin (A \cap B)\} \\ &= \{2,3,4,\ldots, 10,11,12\} \end{aligned}$$

We have summarized the definitions of the various set operations below for your easy reference:

TABLE-1: IMPORTANT SET OPERATIONS	
Union of Sets:	$A \cup B = \{x \mid x \in A \text{ or } x \in B\}$
Intersection of Sets:	$A \cap B = \{x \mid x \in A \text{ and } x \in B\}$
Difference of Sets:	$A - B = \{x \mid x \in A \text{ and } x \notin B\}$
Complement of a Set:	$A' = \{x \mid x \in U \text{ and } x \notin A\}$

EXERCISE SET #4

PART A

Given the sets:

$$A = \{a,b,c,d\}$$
$$B = \{b,c\}$$
$$C = \{c,d,e\}$$
$$U = \{a,b,c,d,e,f\}$$

determine the following:

1. $A \cup B$
2. $B \cup C$
3. A'
4. $A - B$
5. $A - C$
6. $A \cap (B \cup C)$
7. $(A \cap B) \cup (A \cap C)$
8. $(A \cap B)'$
9. $A' \cup B'$
10. $C - B$
11. $B \cup C$
12. $A \cap U$
13. $A - U$
14. $U - A$
15. $(A \cup B)'$
16. $A' \cap B'$

In Exercises 17–23, determine whether each statement is true or false.

17. For any sets A and B, $(A - B) \subseteq B$.
18. For any set A, $A - \emptyset = A$.
19. If A and B are disjoint, then $A \cup B = \emptyset$.
20. If A and B are disjoint, then $\#(A \cup B) = \#A + \#B$.
21. For any sets A and B, $(A \cup B) \subseteq A$
22. For any set A, $A - A = \emptyset$.
23. For any sets A and B, $(A \cap B) \subseteq A$.

PART B

24. Given the sets:

$$A = \{2,1,3,d\},\ B = \{1,d\},\ \text{and}\ U = \{1,2,3,d,e\}$$

Verify each of **DeMorgan's Laws**:

(a.) $(A \cap B)' = A' \cup B'$

(b.) $(A \cup B)' = A' \cap B'$

25. Using the same sets as in Part A above, verify that:

$$A \cap (B \cup C) = (A \cap B) \cup (A \cap C)$$

This is an example of a **distributive law** for sets that can be proved to hold true for any sets A, B, and C.

5 VENN DIAGRAMS

In this lesson, we will learn about what are commonly called **Venn diagrams.** Named in honor of the English mathematician John Venn (1834–1923), these diagrams are a useful way of visualizing the possible relationships that can exist between various sets. Perhaps, the simplest Venn diagram that we can draw is the one for a single set:

In this figure, we represent the set A by a circle, and the universal set U by a rectangle. Every point which lies inside the circle can represent a possible element of A, while every point inside the entire rectangle is viewed as an element of the universal set; this makes good sense since the universal set contains all of the possible elements we need to handle any given problem involving sets.

In order of increasing complexity, the next Venn diagram we shall consider is the one which represents the complement of a set. In Figure-2, the set of points which lie *outside* the circle representing A have been shaded; any point inside this shaded region is an element of the complement of A.

Similarly, we can use Venn diagrams to represent the intersection of two sets as well. Take a look at the next diagram:

Here, we have only shaded the region which is *common* to both A and B since the intersection of two sets consists of those elements common to both sets. Finally, the union of two sets may be depicted as follows:

In Figure-4 we note that the regions that are interior to either A or B have been shaded since the union of two sets is the collection of all of those

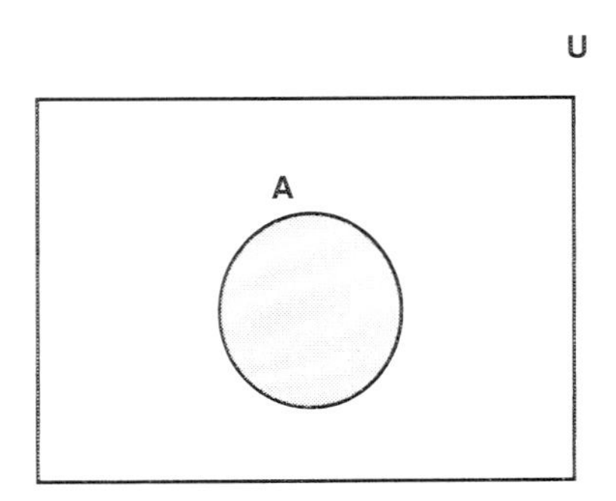

Figure 1. The Venn Diagram for a Single Set A

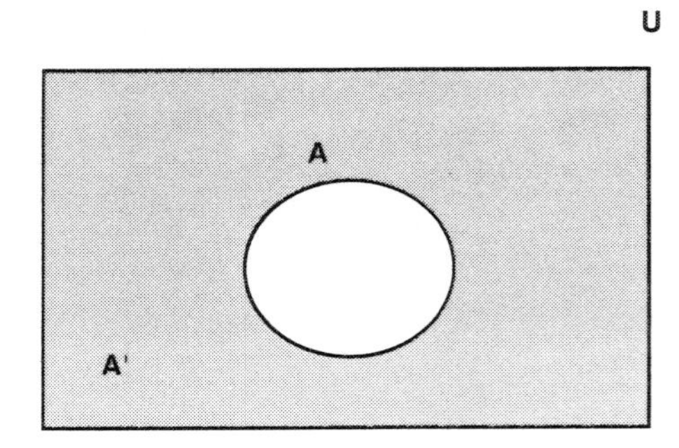

Figure 2. The Complement of the Set A

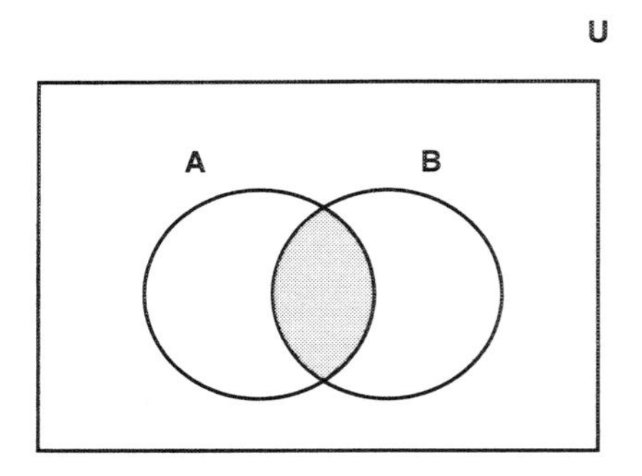

Figure 3. The Intersection of the Sets A and B

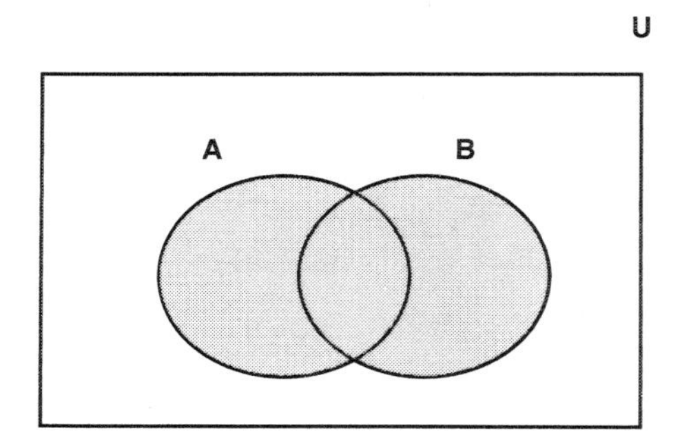

Figure 4. The Union of the Sets A and B

objects that are either in A, in B, or in both. After studying the above examples, we conclude that for the case of two sets, the general situation must look something like the situation depicted in Figure-5.

Each region, or combination of regions in this figure has a precise mathematical meaning, and a summary of those mathematical interpretations appears in the accompanying table:

This table may look terribly complicated at first glance, but a simple example will help the reader to understand these ideas.

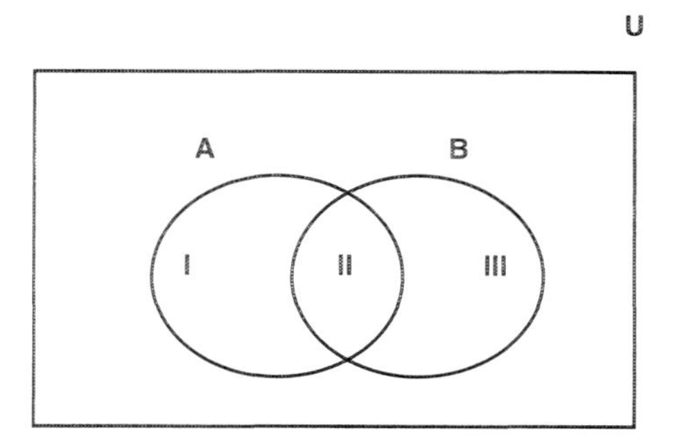

Figure 5. The Possible Regions of a Simple Venn Diagram

Regions of the Venn Diagram	Mathematical Interpretation
I	$A - B$
I and II	A
I, II, and III	$A \cup B$
I, III, and IV	$(A \cap B)'$
I and IV	B'
II	$A \cap B$
II and III	B
III	$B - A$
III and IV	A'
IV	$(A \cup B)'$

Example-1: Given the sets:

$$U = \{a,b,c,d,e,g,h,i,j,k\}$$

$$A = \{a,b,c,d,e\} \text{ and } B = \{c,d,e,g,h\}$$

(a) Draw a Venn diagram which illustrates the relationship between these sets.
(b) What region of the diagram corresponds to $A \cap B$?
(c) What regions represent $A \cup B$?
(d) What region represents $A - B$?

Solution:

(a) We construct the Venn diagram as follows:

Here, we have taken the liberty of using different colors to represent the various regions. Observe that the elements $i,j,$ and k neither belong to A nor B so they have been placed *outside* both of the circles representing these sets.

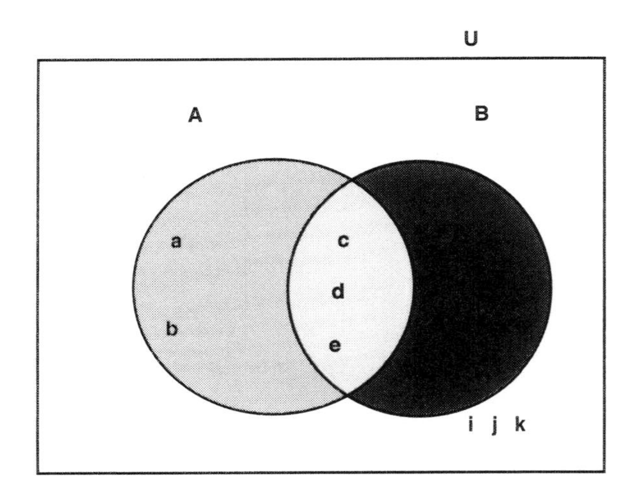

Venn Diagram for Example-1

(b) By direct calculation,

$$A \cap B = \{a,b,c,d,e\} \cap \{c,d,e,g,h\} = \{c,d,e\}$$

and so, we see that the yellow region of the diagram corresponds to the intersection of these two sets.

(c) Similarly, we have:

$$A \cup B = \{a,b,c,d,e\} \cup \{c,d,e,g,h\} = \{a,b,c,d,e,g,h\}$$

and we conclude the union of these sets is represented by the gray, yellow, and light blue regions combined.

(d) We now calculate the difference $A - B$:

$$A - B = \{a,b,c,d,e\} - \{c,d,e,g,h\} = \{a,b\}$$

and we see that their difference corresponds to the gray region of the Venn diagram. Similarly, it's easy to verify that the light blue region represents the set $B - A$.

EXERCISE SET #5

PART A

1. Draw a Venn diagram which illustrates the relationships between the following sets:

$$A = \{1,2,3,4,5\}$$
$$B = \{3,4,5,6,7\}$$
$$U = \{1,2,3,...,9,10\}$$

2. Draw a Venn diagram which illustrates the following sets:

$$A = \{r,s,t,u\}$$
$$B = \{t,u,v,w\}$$
$$U = \{a,c,r,s,t,u,v,w,x,y\}$$

3. Draw a Venn diagram which illustrates the following sets:

$$A = \{2,4,6,8\}$$
$$B = \{1,3,5,7,9\}$$
$$U = \{1,2,3,...,9,10\}$$

4. Draw a Venn Diagram for the following sets of famous mathematicians:

M = {Euler, Gauss, Euclid}
P = {Einstein, Pythagoras, Gauss}
U = {Euler, Gauss, Pythagoras, Euclid, Einstein, Laplace}

5. Draw a Venn Diagram for the following sets:

A = {hearts, clubs, diamonds}
B = {spades, clubs}
U = {clubs, hearts, spades, diamonds, jokers}

In Exercises 6–17, use the Venn Diagram in Figure-6 to determine each of the following sets:

6. A
7. B
8. U
9. $A \cap B$
10. $A \cup B$
11. $A - B$
12. $B - A$

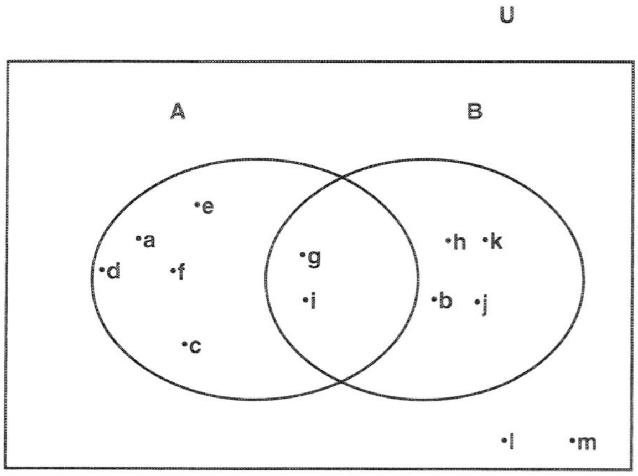

U
A
B
•e
•a
•d
•f
•c
•g
•i
•h •k
•b •j
•l
•m

Figure 6

13. A'
14. B'
15. $A' \cup B'$
16. $(A \cap B)'$
17. $(A \cup B)'$

In Exercises 18–29, use the Venn Diagram in Figure-7 to determine each of the following sets:

U
A
B
•3
•6
•1
•2
•7
•4
•5
•8
•9

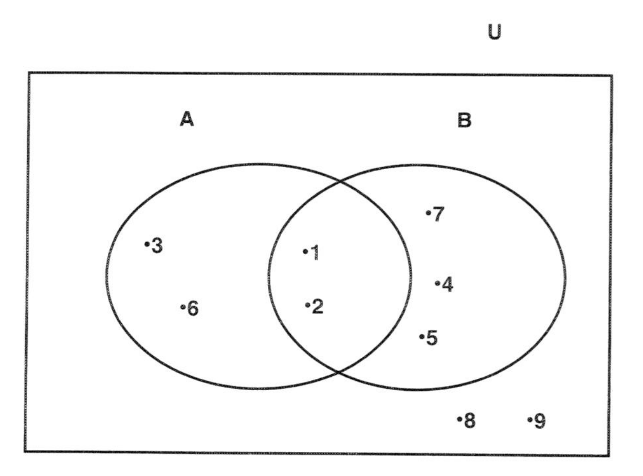

Figure 7

18. A
19. B
20. U
21. $A \cap B$

22. $A \cup B$
23. $A - B$
24. $B - A$
25. A'
26. B'
27. $A' \cup B'$
28. $(A \cap B)'$
29. $(A \cup B)'$

PART B

30. A "generic Venn Diagram" for two sets A and B is given below:

Exercise-30

By appropriately shading this diagram, show that for any two sets A and B:

(a) $(A \cap B)' = A' \cup B'$
(b) $(A \cup B)' = A' \cap B'$

31. By using a Venn diagram similar to that in Exercise-30, show that for any two sets A and B:

$$\#(A \cup B) = \#(A) + \#(B) - \#(A \cap B)$$

32. Use an appropriate Venn diagram to show that for any set A:

$$(A')' = A$$

33. Use the results of Exercise-30, and Exercise-32 to show that for any two sets A and B:

$$(A \cap B')' = A' \cup B$$

6 Applications of Sets

We can use some basic set theory and Venn diagrams to solve many types of problems that occur in everyday life. The best way to understand how this works is to see some simple examples.

Example-1: A local radio station conducted a telephone survey of 100 of its listeners. The survey results indicated that:

50 listeners enjoy Blues music.
60 listeners enjoy Classic Rock music.
20 listeners enjoy both Blues and Classic Rock music.

(a) How many listeners enjoy Blues music but do not enjoy Classic Rock?
(b) How many listeners enjoy Classic Rock but do not enjoy Blues music?
(c) How many listeners enjoy Blues music or Classic Rock?
(d) How many listeners enjoy neither Blues music nor Classic Rock?

Solution:

Step #1: We let B denote the set of people who enjoy Blues music, and let C represent the set of those listeners who enjoy Classic Rock. This information may now be put into mathematical form:

50 listeners enjoy Blues music means $\#(B) = 50$.
60 listeners enjoy Classic Rock music means $\#(C) = 60$.
20 listeners enjoy both Blues and Classic Rock so that $\#(B \cap C) = 20$.

In the steps that follow, we shall now use this information to construct a Venn diagram for the problem:

Step #2: We begin in the *center of the diagram* – this is usually advisable – and fill in the number of elements in region II. Since region II represents $B \cap C$, and we already know that $\#(B \cap C) = 20$, we conclude this region must contain 20 listeners.

Step #3: We now determine the number of listeners that fall into region I. Since 50 listeners enjoy the Blues, but 20 of them also enjoy Classic Rock, we see that this region must contain a total of $50 - 20$ or 30 listeners.

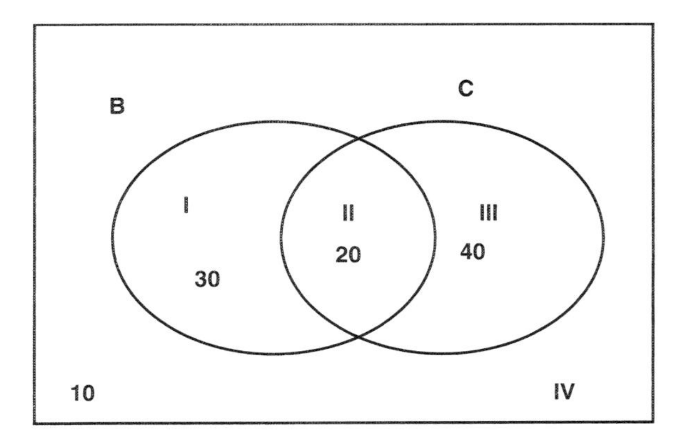

Example-1: Radio Station Survey

Step #4: We shall determine the number of listeners that fall into region III of the Venn diagram. Since 60 people enjoy Classic Rock, but 20 of them also enjoy the Blues, we conclude that region III must contain $60 - 20 = 40$ listeners.

Step #5: Finally, we determine the number of listeners that fall into region IV. Since the number of listeners that enjoy either Blues music or Classic Rock is:

$$30 + 20 + 40 = 90$$

and since a total of 100 listeners were surveyed, we see that region IV must contain a total of $100 - 90$ or 10 listeners. Since we have filled in all of the regions, we are now prepared to answer each of the questions originally posed:

(a) The 30 listeners in region I are those that enjoy Blues music but do not enjoy Classic Rock.
(b) As can be seen from examining region II, there are 40 listeners that enjoy Classic Rock but do not enjoy the Blues.
(c) The number of listeners that enjoy Blues music *or* Classic Rock is represented by $\#(B \cup C)$ or a combination of regions I, II, and III. We see that there is a total of 90 such people.
(d) Finally, region IV represents those listeners who enjoy neither Blues music nor Classic Rock. Examining the diagram, we see that there are 10 listeners who fall into this category.

Remark:
Strictly speaking, in answering question (c) above, we really didn't need to draw a Venn diagram. For this question, we were trying to determine $\#(B \cup C)$. After a little reflection, it would appear reasonable that:

$$\#(B \cup C) = \#(B) + \#(C) - \#(B \cap C) \quad (1)$$

where we "subtract off" the number of elements in the intersection of these two sets since if an element is common to both B and C then it is only listed *once* in the union of these sets. In fact, for the information given in Example-1 we have:

$$\#(B \cup C) = 90$$

and

$$\#(B) + \#(C) - \#(B \cap C) = 50 + 60 - 20 = 90$$

so we see that our conjecture has been verified. Mathematicians call equations such as (1) a **counting formula,** and it may be shown that for any *finite* sets A and B:

COUNTING FORMULA

$$\#(A \cup B) = \#(A) + \#(B) - \#(A \cap B) \quad (2)$$

This formula is sometimes useful in reducing and checking our work when our Venn diagrams involve only two sets.

Example-2: Gracie's Bistro, a fine dining establishment located in Cranston, Rhode Island, surveyed a total of 108 of its customers over a period of several days. The survey results showed that:

83 patrons thought the service was excellent.
92 patrons reported the food was of excellent quality.
72 patrons felt that both the service and food were excellent.

(a) How many customers thought they received excellent service but the food was not of an excellent quality?
(b) How many patrons felt that the food was of excellent quality but the service was not excellent?
(c) How many customers reported either the service *or* the food to be excellent?
(d) How many patrons felt that neither the service nor the food was excellent?

Step #1: We let S denote the set of people who felt that the service they received was excellent, and let Q represent the set of those patrons who felt that the food was of excellent quality. As in the previous example, we now put this information into mathematical form:

83 patrons thought the service was excellent means that:

$$\#(S) = 83$$

92 patrons reported the food was of excellent quality, so

$$\#(Q) = 92$$

72 patrons felt that both the service and food were excellent implies that:

$$\#(S \cap Q) = 72$$

As before, in the steps that follow, we now use this information to construct a Venn diagram for the problem:

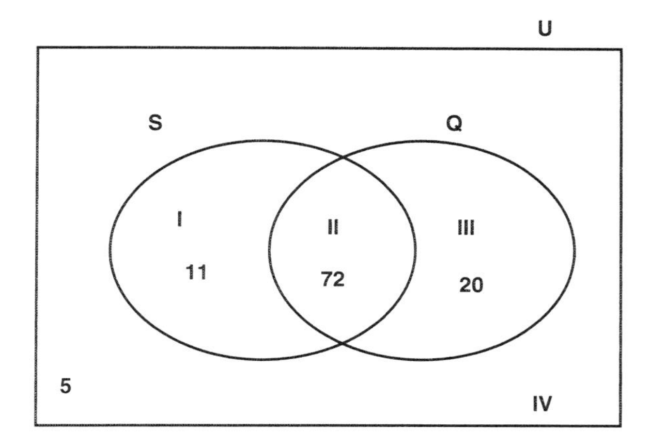

Example-2: Gracie's Bistro

Step #2: Once again, we begin in the *center of the diagram* and fill in the number of elements in region II. Since region II represents $S \cap Q$, and from above, we already know that $\#(S \cap Q) = 72$, we conclude this region must contain 72 patrons.

Step #3: We now determine the number of patrons that fall into region I. Since 83 patrons felt the service was excellent, but 72 of them also felt the quality of food was excellent, we are forced to conclude that this region contains a total of $83 - 72$ or 11 patrons.

Step #4: We shall now determine the number of patrons that fall into region III of the Venn diagram. Since 92 people felt the food quality was excellent, but 72 of them also reported excellent service, we conclude that region III must contain $92 - 72$, or 20 patrons.

Step #5: Finally, we determine the number of patrons that fall into region IV. Since the number of patrons who feel that either service or food quality was excellent is:

$$11 + 72 + 20 = 103$$

and a total of 108 patrons were surveyed, we must conclude that region IV contains a total of $108 - 103$ or just 5 patrons. As in the previous example, since we have completed the process of filling in all of the regions in the Venn diagram, we are now in a position to answer each of the questions originally posed:

(a) The 11 patrons in region I are those that who reported excellent service but did not feel the food quality was excellent.
(b) From region II, we conclude there are 72 patrons who felt that both food quality and service were excellent.
(c) The number of patrons who reported that either the service or food quality were excellent is given by $\#(S \cup Q)$, which is a combination of regions I, II, and III. From the diagram, we see that there a total of 103 such people. Another way of obtaining the same answer would be to use the counting formula:

$$\begin{aligned}\#(S \cup Q) &= \#(S) + \#(Q) - \#(S \cap Q)\\ &= 83 + 92 - 72\\ &= 103\end{aligned}$$

which agrees with our previous answer.

(d) Finally, region IV represents those patrons who felt that neither the service nor food were of excellent quality. Examining the diagram, we see that there 5 patrons who fall into this group.

Example-3: In order to improve customer service, the manager of the Fly-by-Night Travel Agency personally interviewed 101 people to determine which modes of the transportation they enjoyed. The manager determined that:

44 people enjoyed traveling by plane.
46 people enjoyed traveling by train.
26 people enjoyed traveling by bus.
11 enjoyed traveling by both plane and train.
9 enjoyed traveling by both plane and bus.
13 enjoyed traveling by bus and train.
6 enjoyed traveling by bus, train, and plane.

How may customers enjoyed:

(a) at least one mode of travel?
(b) only traveling by plane?
(c) traveling by both plane and bus but *not* by train?
(d) none of the modes of travel?

Solution:

Step #1: We let P denote the set of people who enjoy travelling by plane, let T represent the set of those people who like to travel by train, and let B

be the set of people who enjoy traveling by bus. As in the above examples, we now be put this information into mathematical form. Due to volume of information in this example, however, we shall list this information in the following table:

Given Information	Mathematical Form
44 people enjoyed to traveling by plane	$\#(P) = 44$
46 people enjoyed traveling by train	$\#(T) = 46$
26 people enjoyed traveling by bus	$\#(B) = 26$
11 enjoyed traveling by both plane and train	$\#(P \cap T) = 11$
9 enjoyed traveling by both plane and bus	$\#(P \cap B) = 9$
13 enjoyed traveling by bus and train.	$\#(B \cap T) = 13$
6 enjoyed traveling by bus, train, and plane	$\#(B \cap T \cap P) = 6$

As before, in the steps that follow, we now use this information to construct a Venn diagram for the problem:

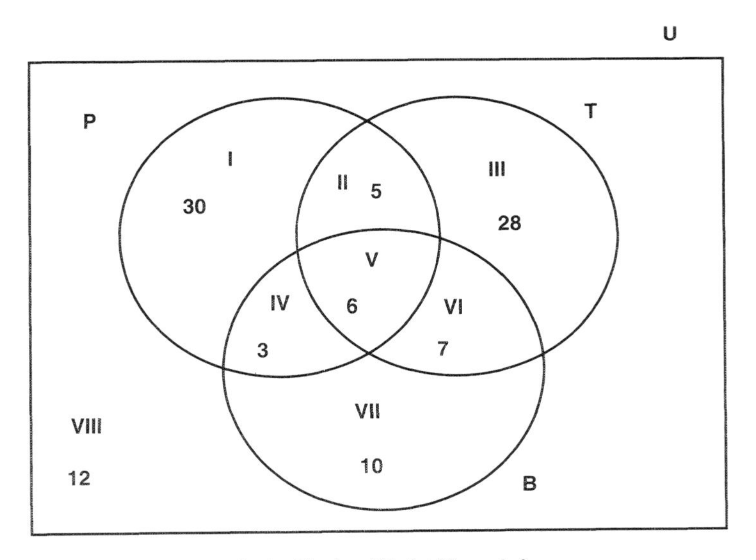

Example-3: Fly-by-Night Travel Agency

Step #2: As in the past, we begin our analysis in the very center of the Venn diagram, or region V. This region is represented by: $\#(P \cap T \cap B)$ and from the table, it must contain 6 people.

Step #3: We now determine the number of elements in region II. From the diagram, we note that regions II and V combined represent the set $P \cap T$, and from the table, we know that this set contains 11 elements. Consequently, the number of people who fall into region II must be $11 - 6$ or exactly 5 elements.

Step #4: To determine the number of elements in region IV, we observe that both regions IV and V combined represent $P \cap B$, and since we are given that $\#(P \cap B) = 9$, we conclude that region IV must contain $9 - 6 = 3$ elements.

Step #5: To determine how many people fall into region VI, we observe that if we combine region VI with region V we get the set $B \cap T$, and from the above table $\#(B \cap T) = 13$. Consequently, region VI must contain $13 - 6 = 7$ elements.

Step #6: To determine the number of people that fall into Region I we note that when regions I, II, IV, and V are combined, they make up the set P. Since $\#(P) = 44$, region II has 5 elements, region IV has 3 elements, and region V has 6 elements, then we may conclude that region I contains:

$$44 - (5 + 3 + 6) = 30$$

or 30 elements. Similar reasoning [Try it!] will show us that region III contains 28 elements, and region VII contains 10 elements.

Step #7: Finally, we determine the number of elements in region VIII. Since this region consists of those elements that do not fall into any of the three sets, and total of 101 people were surveyed, region VIII must contain a total of:

$$101 - (30 + 5 + 28 + 3 + 6 + 7 + 10) = 12$$

12 elements. Now that all of the regions in the Venn diagram have been accounted for, we can answer each of the above questions.

(a) The number of people who enjoy at least one mode of travel is represented by $\#(P \cup B \cup T)$ or the number of elements in regions I, II, III, IV, V, VI, and VII combined. Thus, there are

$$30 + 5 + 28 + 3 + 6 + 7 + 10 = 89$$

people who enjoy at least one mode of travel.

(b) The number of people who only enjoy traveling by plane is by region I. We have already determined this to be 30.

(c) The number of people who enjoy traveling by both plane and bus but *not* by train is the number of elements in region III. From the Venn diagram, we conclude there are only 3 such people in this group.

(d) The number of people who don't enjoy any given mode of travel is determined from region VIII. We have already found that there are exactly 12 people in this category.

EXERCISE SET #6

PART A

1. A classical radio station conducted a survey of 120 of its listeners to determine their favorite composers. The survey results indicated that:

   60 listeners like Bach.
   50 listeners like Mozart.
   30 listeners like both Bach and Mozart.

   Of those people surveyed, how many:

   (a) like Bach?
   (b) like Mozart?
   (c) like Bach but don't like Mozart?
   (d) like Mozart but don't like Bach?
   (e) like neither Bach nor Mozart?

2. A restaurant surveyed 80 of its customers to determine whether they liked to the use of butter or margarine. The survey results were as follows:

   22 customers liked butter.
   38 customers liked margarine.
   12 customers liked both butter and margarine.

   Of those customers surveyed, how many:

   (a) like butter?
   (b) like margarine?
   (c) like butter but don't like margarine?
   (d) like margarine but don't like butter?
   (e) like neither butter nor margarine?

3. A total of 1,000 students at Johnson & Wales University were surveyed to determine their course-scheduling preferences. The survey results were as follows:

   620 students liked morning classes
   320 students liked afternoon classes
   230 students liked both morning and afternoon classes.

   Of those students surveyed, how many:

   (a) like morning classes?
   (b) like afternoon classes?
   (c) like morning classes but don't like afternoon classes?

(d) like afternoon classes but don't like morning classes?
(e) like neither morning nor afternoon classes?

4. The Crown Hotel surveyed 150 of its guests to determine their level of satisfaction. The survey indicated that:

   120 guests liked the hotel's food.
   100 guests liked the room accomodations.
   80 guests liked both the hotel's food and the room accomodations.

   Of those guests surveyed, how many:

   (a) like the hotel's food?
   (b) like the room accomodations?
   (c) like the food but don't like the room?
   (d) like the room but not the food?
   (e) like neither the room nor the food?

5. A total of 250 patients at the St. James Hospital were interviewed to determine their degree of satisfaction with hospital care. The survey results indicated that:

   220 patients felt they received quality care.
   180 patients felt the hospital staff was courteous.
   170 patients felt they received quality care, and thought the staff was courteous.

   Of those patients surveyed, how many:

   (a) felt they received quality care?
   (b) felt the hospital staff was courteous?
   (c) felt they received quality care but didn't feel that the hospital's staff was courteous?
   (d) liked the courtesy of the hospital's staff but didn't feel that they received quality care?
   (e) didn't feel they received quality care and didn't feel the hospital's staff was courteous?

6. The Dean's Office surveyed 250 students to determine their course preferences. The survey results were as follows:

   65 students liked Mathematics courses.
   45 students liked English courses.
   30 students liked History courses.
   15 students liked Mathematics and English.
   12 students liked Mathematics and History.
   21 students liked English and History.
   11 students liked all three subjects.

How many students:

(a) only liked Mathematics?

(b) only liked English?
(c) only liked History?
(d) liked Mathematics but not English?
(e) liked History but not Mathematics?
(f) didn't like any of the subjects?

7. In order to improve customer service, the manager of a retail store survey 400 customers to determine their methods of payment. The manager determined that:

   120 customers liked to use a VISA card.
   225 customers liked to use a MASTER card.
   20 customers liked to pay by cash.
   11 customers payed by VISA card and cash.
   10 customers payed by MASTER card and cash.
   12 customers payed by MASTER card and VISA card.
   8 customers payed by all three methods.

How many customers used:

(a) at least one method of payment?
(b) only a VISA card?
(c) only a MASTER card?
(d) only made payments in cash?
(e) used a VISA card or cash but not a MASTER card?
(f) used a MASTER card or cash but not a VISA card?
(g) used a VISA card or a MASTER card but not cash?
(h) didn't use any of the above methods of payment?

7 Introduction to Equations

We now turn our attention to the concept of an equation and to some of the simple methods used to solve equations. An **equation** is a statement which *asserts* that two quantities are equal. Like any other statement that someone may make, an equation may be either true or false, but not both. Take a look at the following equations:

$$2 + 2 = 3 + 1$$
$$\frac{4}{2} + 1 = \frac{8}{4} + 2$$

The first equation is certainly true since the *lefthand side* of the equation $(2 + 2 = 4)$ is equal to the *righthand side* $(3 + 1 = 4)$ of the equation. The second equation is false since the lefthand and righthand sides of this equation represent different numbers:

$$\frac{4}{2} + 1 = 3 \text{ but } \frac{8}{4} + 2 = 4$$

An **algebraic equation** is an equation which contains one or more variables, and there are two kinds of algebraic equations: **identities,** and **conditional equations.** An **identity** is an algebraic equation which is true for *all* values of its variables. Consider the following statements:

$$x + y = y + x$$
$$2(x + y) = 2x + 2y$$

It is clear that both of these equations are true for *all* real numbers x and y; in fact, the first equation is just the commutative law for addition, and the second equation follows immediately from a simple application of the distributive law. Consequently, we conclude that both of these equations are identities.

On the other hand, a **conditional equation** is one which is only true for *some* value(s) of its variables. The following equations are examples of conditional equations:

$$2x + 1 = x + 2 \tag{1}$$

$$x^2 - 4 = 0 \tag{2}$$

The first equation is true when $x = 1$ since if we substitute this value of x into the equation we get a true statement:

$$2(1) + 1 = 3 = (1) + 2 \quad \text{[Substitute } x = 1 \text{ into (1)]}$$

Stated differently, we say that $x = 1$ is a **solution** of Equation (1) because when this value of x is substituted into Equation (1), we obtain a true statement.

Similarly, Equation (2) is true when $x = -2$ or when $x = 2$. In order to verify this assertion, we substitute these values of x into Equation (2) to see if we get a true statement:

$$(-2)^2 - 4 = 4 - 4 = 0 \quad \text{[Substitute } x = -2 \text{ into (2)]}$$

$$(2)^2 - 4 = 4 - 4 = 0 \quad \text{[Substitute } x = 2 \text{ into (2)]}$$

As both of these statements are true, we conclude that $x = -2$ and $x = 2$ are solutions of Equation (2).

A question which naturally arises is "How do we solve conditional equations?" As it turns out, certain types of equations are very easy to solve while others are quite difficult. In this lesson, we shall examine the kind of equations that fall under the very easy category: linear equations in one variable.

Definition-1: (Linear Equation in One Variable)

A **linear equation** (in one variable) is a conditional equation which can be written in the form:

$$ax + b = cx + d \tag{3}$$

where $a, b, c,$ and d are real numbers and not both a and c are zero.

Example-1: Determine whether each of the following equations is a linear equation.

(a) $x + 1 = 2$
(b) $x + 7 = 2x - 1$
(c) $4x = 0$
(d) $x^2 + 2x = x + 2$

Solution:

(a) If we start with Equation (3), and let:

$$a = 1, b = 1, c = 0, \text{ and } d = 2$$

we get:

$$(1)x + (1) = (0)x + (2)$$
$$\text{or } x + 1 = 2$$

So we conclude that this equation is a linear equation.
(b) Once again, if we start with Equation (3), and let:

$$a = 1, b = 7, c = 2, \text{ and } d = -1$$

we get:

$$(1)x + (7) = (2)x + (-1)$$
$$\text{or } x + 7 = 2x - 1$$

So we conclude that this equation is a linear equation as well.
(c) Similarly, using Equation (3), if we let:

$$a = 4, b = 0, c = 0, \text{ and } d = 0$$

we obtain:

$$(4)x + (0) = (0)x + (0)$$
$$\text{or } 4x = 0$$

So we conclude that the third equation is a linear equation.
(d) The equation $x^2 + 2x = x + 2$ is not a linear one since x appears to the second power on the lefthand side of the equation.

After working through this example, the reader has probably concluded that in order for an equation to be linear, the unknown variable [x in Equation (3)] must always appear to the *first power.* This observation is, in fact, always valid.

In order to solve linear equations, we usually use what are commonly called the "Properties of Equalities", and use these properties to determine the unknown variable; so let's examine these useful properties.

Theorem-1: (Addition/Subtraction Property)

If we are given the true equation:

$$a = b$$

then the following equations are also true for any real number c:

$$a + c = b + c \tag{4}$$

$$a - c = b - c \tag{5}$$

In other words, *the same number may be added (or subtracted) from both sides of a true equation, and the result is a true equation.*

Solution:

(a) The basic idea in solving all linear equations is to *isolate* the unknown variable x. In this case, we see that we can "get x by itself" if we add 4 to both sides of this equation:

$$
\begin{aligned}
x - 4 &= 10 \\
x - 4 + 4 &= 10 + 4 && \text{[Add 4 to both sides.]} \\
x + 0 &= 14 && \text{[Simplify]} \\
x &= 14 && \text{Solution.}
\end{aligned}
$$

(b) Similarly, to isolate x in this problem we subtract 11 from both sides of the equation to get:

$$
\begin{aligned}
x + 11 &= 17 \\
x + 11 - 11 &= 17 - 11 && \text{[Subtract 11 from both sides.]} \\
x + 0 &= 6 && \text{[Simplify]} \\
x &= 6 && \text{Solution.}
\end{aligned}
$$

(c) Here, we subtract 2π from both sides of the equation in order to isolate the unknown variable x:

$$
\begin{aligned}
x + 2\pi &= 0 \\
x + 2\pi - 2\pi &= 0 - 2\pi && \text{[Subtract } 2\pi \text{ from both sides.]} \\
x + 0 &= -2\pi && \text{[Simplify]} \\
x &= -2\pi && \text{Solution.}
\end{aligned}
$$

(d) Finally, we have:

$$
\begin{aligned}
x - 21 &= 34 \\
x - 21 + 21 &= 34 + 21 && \text{[Add 21 to both sides.]} \\
x + 0 &= 55 && \text{[Simplify]} \\
x &= 55 && \text{Solution.}
\end{aligned}
$$

Theorem-2: (Multiplication/Division Property)

If we are given the true equation:

$$a = b$$

then the following equations are also true for any non-zero real number c:

$$a \cdot c = b \cdot c \tag{6}$$

$$a \div c = b \div c \tag{7}$$

In other words, *if both sides of a true equation are multiplied (or divided) by the same non-zero number, then the resulting equation is also true.*

(d) Finally, we have:

$$\begin{aligned} x - 21 &= 34 \\ x - 21 + 21 &= 34 + 21 && \text{[Add 21 to both sides.]} \\ x + 0 &= 55 && \text{[Simplify]} \\ x &= 55 && \text{Solution.} \end{aligned}$$

Theorem-2: (Multiplication/Division Property)

If we are given the true equation:

$$a = b$$

then the following equations are also true for any non-zero real number c:

$$a \cdot c = b \cdot c \tag{6}$$

$$a \div c = b \div c \tag{7}$$

In other words, *if both sides of a true equation are multiplied (or divided) by the same non-zero number, then the resulting equation is also true.*

Example-3: Solve each of the following equations:

(a) $2x = 14$

(b) $-\frac{1}{3}x = 7$

(c) $\frac{2}{3}x = 12$

(d) $9x = -81$

Solution:

(a) Using Theorem-2, we can isolate x by dividing both sides of the equation by 2 as follows:

$$\begin{aligned} 2x &= 14 \\ \frac{2x}{2} &= \frac{14}{2} && \text{[Divide both sides by 2.]} \\ 1 \cdot x &= 7 && \text{[Simplify]} \\ x &= 7 && \text{Solution.} \end{aligned}$$

Note that we could have achieved the same result by *multiplying* both sides of the equation by the **multiplicative inverse** of 2:

$$\begin{aligned}
2x &= 14 \\
\frac{1}{2}\cdot(2x) &= \frac{1}{2}\cdot(14) && \text{[Multiply both sides by } 1/2.] \\
1\cdot x &= 7 && \text{[Simplify]} \\
x &= 7 && \text{Solution.}
\end{aligned}$$

Whether we choose to divide or to multiply doesn't matter as long as we play by the rules!
(b) Similarly,

$$\begin{aligned}
-\frac{1}{3}x &= 7 \\
-3\cdot(-\frac{1}{3}x) &= -3\cdot 7 && \text{[Multiply both sides by } -3.] \\
1\cdot x &= -21 && \text{[Simplify]} \\
x &= -21 && \text{Solution.}
\end{aligned}$$

Notice that we multiplied both sides of this equation by the multiplicative inverse of $-1/3$. We could have *divided* both sides of the equation by $-1/3$ and arrived at the same result, but our approach was less messy.
(c) Here, in order to isolate x on the lefthand side of the equation, we multiply both sides of the equation by the multiplicative inverse of $2/3$:

$$\begin{aligned}
\frac{2}{3}x &= 12 \\
\frac{3}{2}\cdot(\frac{2}{3}x) &= \frac{3}{2}\cdot(12) && \text{[Multiply both sides by 3/2.]} \\
1\cdot x &= 18 && \text{[Simplify]} \\
x &= 18 && \text{Solution.}
\end{aligned}$$

(d) Similarly, we obtain:

$$\begin{aligned}
9x &= -81 \\
\frac{9x}{9} &= \frac{-81}{9} && \text{[Divide both sides by 9.]} \\
1\cdot x &= -9 && \text{[Be careful with signs.]} \\
x &= -9 && \text{Solution.}
\end{aligned}$$

We are now in a position to combine the ideas in Theorem-1 and Theorem-2 to solve more complicated types of linear equations. The basic method for solving any linear equation – no matter how complicated – is as follows:

GENERAL METHOD FOR SOLVING LINEAR EQUATIONS

In order to solve a general linear equation such as:

$$ax + b = cx + d$$

where $a,b,c,$ and d are real numbers, we use the following procedure:

Step #1: We use the Addition/Subtraction Property (repeatedly if necessary) to put all of the terms which involve the unknown variable x on one side of the equation, and all of the constants (numbers) on the other side of the equation.

Step #2: Whenever possible, on an ongoing basis, we continually simplify each side of the equation by combining *like terms* which involve x, and by combining constants.

Step #3: At this point, we should have obtained an equation that looks like:

$$Ax = B$$

where A and B are just real numbers. We now use the Multiplication/Division Property on this equation in order to finally isolate the unknown variable x.

Although this general rule may seem vague at first, some concrete examples will clarify this procedure.

Example-4: Solve each of the following linear equations:

(a) $3x + 1 = x - 4$

(b) $-x + 10 = 2x + 20$

(c) $\frac{1}{2}x + 3 = \frac{1}{3}x + 4$

(d) $7(x + 1) = 2x + 5$

Solution:

(a) We will *repeatedly* use the Addition/Subtraction Property to move all of the x's to the lefthand side of the equation, and move all of the constants to the righthand side of the equation. Observe that we will simplify as we move along:

$$3x + 1 = x - 4$$
$$3x + 1 - x = x - 4 - x \quad \text{[Subtract } x \text{ from both sides.]}$$
$$2x + 1 = -4 \quad \text{[Combine like terms.]}$$
$$2x + 1 - 1 = -4 - 1 \quad \text{[Subtract 1 from both sides.]}$$
$$2x = -5 \quad \text{[Simplify]}$$

We have now obtained an equation that looks like $Ax = B$, so we use the Multiplication/Division Property to isolate x:

$$\frac{2x}{2} = \frac{-5}{2} \quad \text{[Divide both sides by2.]}$$
$$x = -\frac{5}{2} \quad \text{Solution.}$$

(b) Once again, we will *repeatedly* use the Addition/Subtraction Property to move all of the x' s to one side of the equation, and move all of the constants to the other side:

$$-x + 10 = 2x + 20$$
$$-x + 10 - 10 = 2x + 20 - 10 \quad \text{[Subtract 10 from both sides.]}$$
$$-x = 2x + 10 \quad \text{[Simplify]}$$
$$-x - 2x = 2x + 10 - 2x \quad \text{[Subtract } 2x \text{ from both sides.]}$$
$$-3x = 10 \quad \text{[Simplify]}$$
$$\frac{-3x}{-3} = \frac{10}{-3} \quad \text{[Divide both sides by } -3.\text{]}$$
$$x = -\frac{10}{3} \quad \text{Solution.}$$

The procedure for solving (c) and (d) is similar, and has been listed below in great detail:

(c) Similarly, we have:

$$\frac{1}{2}x + 3 = \frac{1}{3}x + 4$$
$$\frac{1}{2}x + 3 - \frac{1}{3}x = \frac{1}{3}x + 4 - \frac{1}{3}x \quad \text{[Subtract } \frac{1}{3}x \text{ from both sides.]}$$
$$\frac{1}{6}x + 3 = 4 \quad \text{[Combine like terms.]}$$
$$\frac{1}{6}x + 3 - 3 = 4 - 3 \quad \text{[Subtract 3 from both sides.]}$$
$$\frac{1}{6}x = 1 \quad \text{[Simplify]}$$
$$6 \cdot \frac{1}{6}x = 6 \cdot 1 \quad \text{[Multiply both sides by 6.]}$$

$$1 \cdot x = 6 \qquad \text{[Simplify]}$$
$$x = 6 \qquad \text{Solution}$$

(d) And finally,

$$7(x+1) = 2x = 5$$
$$7x + 7 = 2x + 5 \qquad \text{[Simplify using Distributive Law.]}$$
$$7x + 7 - 2x = 2x + 5 - 2x \qquad \text{[Subtract } 2x \text{ from bothsides.]}$$
$$5x + 7 = 5 \qquad \text{[Combine like terms.]}$$
$$5x + 7 - 7 = 5 - 7 \qquad \text{[Subtract 7 from both sides.]}$$
$$5x = -2 \qquad \text{[Simplify]}$$
$$\frac{5x}{5} = \frac{-2}{5} \qquad \text{[Divide both sides by 5.]}$$
$$1 \cdot x = -\frac{-2}{5} \qquad \text{[Simplify]}$$
$$x = -\frac{2}{5} \qquad \text{Solution.}$$

Now that we've had an opportunity to gain some confidence in solving linear equations, we would like to mention one last tip that sometimes helps to simplify linear equations that contain fractions. The basic idea is to multiply both sides of the equation by the **lowest common denominator (LCD)** of all of the fractions that are present; recall that the lowest common denominator of two or more fractions is simply the *smallest* number that the denominators of each fraction divide evenly.

Multiplying both sides of an equation by the lowest common denominator of all of the fractions present "clears the decks" of all fractions, and makes the equation relatively easy to solve. Take a look at the last example of this lesson:

Example-5: Solve the linear equation:

$$\frac{1}{4}x + \frac{2}{3} = \frac{5}{6}x - \frac{1}{2}$$

Solution:

(a) Here, note that the LCD is 12, so we begin as follows:

$$12 \cdot \left(\frac{1}{4}x + \frac{2}{3}\right) = 12 \cdot \left(\frac{5}{6}x - \frac{1}{2}\right) \qquad \text{[Multiply both sides by 12.]}$$
$$\frac{12}{4}x + \frac{24}{3} = \frac{60}{6}x - \frac{12}{2} \qquad \text{[Distributive Law]}$$
$$3x + 8 = 10x - 6 \qquad \text{[Simplify]}$$
$$3x + 8 - 3x = 10x - 6 - 3x \qquad \text{[Subtract } 3x \text{ from both sides.]}$$
$$8 = 7x - 6 \qquad \text{[Simplify]}$$
$$8 + 6 = 7x - 6 + 6 \qquad \text{[Add 6 to both sides.]}$$
$$14 = 7x \qquad \text{[Simplify]}$$
$$\frac{14}{7} = \frac{7x}{7} \qquad \text{[Divide both sides by 7.]}$$
$$2 = x \;\; or \;\; x = 2 \qquad \text{Solution.}$$

EXERCISE SET #7

PART A

In Problems 1–14, Solve each linear equation for the unknown variable x. Check your answers by direct substitution.

1. $x + 4 = 16$
2. $2x + 4 = 10$
3. $3x + 3 = 12$
4. $2x - 7 = 12x + 13$
5. $3(x + 1) = x + 7$
6. $\frac{1}{3}x + 3 = 2x - 2$
7. $17x + 1 = 20x - 2$
8. $10x + 10 = x + 1$
9. $2x - 8 = x - 3$
10. $4x = \frac{2}{5}x + 36$
11. $6x + 2 = 3x - 1$
12. $\frac{1}{2}x + \frac{1}{3} = \frac{1}{3}x - \frac{1}{6}$ [**Hint:** Get rid of all fractions first.]
13. $\frac{1}{4}x = \frac{5}{24}x - 1$
14. $5x - 15 = 10x + 10$

In Exercises 15–21, state whether each of the following statements is true or false:

15. The equation: $x^3 - 2x + 1$ is a linear equation.
16. In a linear equation, the unknown variable always appears to the first power.
17. Given two real numbers a and b such that $a = b$, we may conclude that $a^2 = b^2$.
18. In Theorem-2, we specified that c cannot be zero because division by zero is not defined.
19. The solution of the equation $ax = b$ (where $a \neq 0$) is given by: $x = b/a$.
20. When we combine like terms on any given side of an equation, we are really using the distributive law.

21. The linear equation $2x + 1 = 2x - 3$ has a solution.

PART B

22. By solving the general linear equation:

$$ax + b = cx + d$$

show that its solution is given by:

$$x = \frac{d-b}{a-c} \quad (\text{provided } a \neq c)$$

23. A student was asked to solve the simple equation:

$$x = 2x$$

He then divided both sides of the equation by x to obtain:

$$\frac{x}{x} = \frac{2x}{x}$$

and finally simplified the last result to get:

$$1 = 2$$

which is absurd! What went wrong?

24. Show that the simple linear equation:

$$ax = b$$

has the solution:

$$x = \frac{b}{a} \quad (\text{provided } a \neq 0)$$

25. Prove that if the solution to the linear equation given in Exercise-24 exists, then it is *unique*, i.e., there can only be *one* solution to this problem. **Hint:** Use Proof by Contradiction. Begin by assuming to the contrary, that the equation has two *different* solutions x_1 and x_2 so that we must have both:

$$ax_1 = b \text{ and } ax_2 = b$$

Now try to arrive at a contradiction of your assumption.

26. Solve the following linear equation for x:

$$\frac{x-1}{4} = \frac{2x-1}{7}$$

27. Solve for x:

$$\frac{x+1}{2}=\frac{x+3}{3}$$

28. Solve for x:

$$x+\frac{1}{2}x+\frac{1}{3}x+\frac{1}{4}x=50$$

29. Solve the following linear equation for y:

$$\frac{ay-b}{c}=\frac{by+a}{c}$$

where a,b,c are real constants, and $c \neq 0$.

8 Ratio and Proportion

We define the **ratio** of two quantities a and b as the *quotient* which is obtained when we divide a by b. Consequently, the ratio of a and b may be written as:

$$\frac{a}{b} \quad \text{or} \quad a \div b$$

and a third way of writing the same thing is:

$$a : b \text{ (which is read: "the ratio of } a \text{ to } b.\text{")} \qquad (1)$$

Ratios provide us with a simple method of comparing the relative sizes of two quantities a and b. Take a look at the following simple examples:

Example-1: During a given baseball season, a ballplayer got a total of 120 hits in 400 times at bat. Find the ratio of "hits to times at bat."

Solution:
Since a ratio is just a quotient, we find that the ratio of "hits" to "times at bat" is given by:

$$\frac{120}{400} = \frac{3}{10}$$

So after reducing the initial fraction, we see that the required ratio is 3 : 10, i.e., 3 hits to every 10 times at bat.

Example-2: In a multiple choice history exam consisting of 50 questions, a student got 42 questions correct, and 8 questions incorrect. Find the following ratios:
(a) The ratio of "correct questions" to "incorrect questions."
(b) The ratio of "incorrect questions" to "correct questions."
(c) The ratio of "correct questions" to "total questions answered."

Solution: As in the previous example, we have:
(a) The ratio of "correct questions" to "incorrect questions" is given by:

$$\frac{42}{8} = \frac{21}{4}$$

or 21 : 4.

(b) The ratio of "incorrect questions" to "correct questions" is simply:

$$\frac{8}{42} = \frac{4}{21}$$

or 4 : 21. Note that this is simply the reciprocal of our answer in part (a).
(c) The ratio of "correct questions" to "total questions answered" is given by:

$$\frac{42}{50} = \frac{21}{25}$$

or 21 : 25.

Now, a **proportion** is an equation which asserts that two given ratios are equal. This means that the typical proportion looks like an equation of the form:

GENERAL FORM OF A PROPORTION

$$\frac{a}{b} = \frac{c}{d} \quad \text{(where } b, d \neq 0) \qquad (2)$$

The proportion shown in Equation (2) may be read "a is to b as c is to d." In Equation (2), it is customary to call the quantities a and d the **extremes**, while b and c are called the **means**.

Theorem-1: (Basic Theorem for Proportions)

If the proportion given by:

$$\frac{a}{b} = \frac{c}{d} \quad \text{(where } b, d \neq 0) \qquad (2)$$

is a true statement, then:

$$a \cdot d = b \cdot c \qquad (3)$$

In other words, *if we are given a true proportion, then the product of the means is equal to the product of the extremes.*

A different way to stating Theorem-1 is that if we are given a true proportion, then we may simply **cross multiply** in order to eliminate all of the fractions originally present in the given proportion. Now take a look at the following example:

Example-3: Use Theorem-1 to solve for x in each proportion:

(a)
$$\frac{x}{2}=\frac{30}{40}$$

(b)
$$\frac{3}{8}=\frac{12}{x}$$

(c)
$$\frac{x-1}{3}=\frac{x}{2}$$

Solution: Using Theorem-1 in each case, we obtain:

(a)
$$\frac{x}{2}=\frac{30}{40}$$
$$40x = 2\cdot 30 \quad \text{(Cross Multiply)}$$
$$40x = 60$$
$$x=\frac{60}{40}=\frac{3}{2} \quad \text{Solution}$$

(b)
$$\frac{3}{8}=\frac{12}{x}$$
$$3x = 8\cdot 12 \quad \text{(Cross Multiply)}$$
$$3x = 96$$
$$x=\frac{96}{3}=32 \quad \text{Solution}$$

(c)
$$\frac{x-1}{3}=\frac{x}{2}$$
$$2(x-1)=3x \quad \text{(Cross Multiply)}$$
$$2x-2=3x$$
$$x=-2 \quad \text{Solution}$$

Proportions often arise in the practical applications of mathematics because two variables are often related in a special way: one variable may be **directly proportional** to another variable. This concept is explained in the definition below:

Definition-1: (Direct Variation)

Given two variables x and y, we say that y is **directly proportional** to x, or y **varies directly** as x if and only if:

$$y = kx \tag{4}$$

for some *non-zero* constant k.

Example-4: A Geo Tracker gets 28 miles per gallon of gas.
If x is the number gallons used, and y is the number of miles travelled, then we have:

$$y = 28x$$

so we see that the number of miles traveled is *directly proportional* to the number of gallons of gas used. Note here, that our **constant of proportionality** is $k = 28$.

Example-5: In Jen's Retail Shoe Outlet, the retail price for a certain brand of sneakers is \$75. If n is the number of such sneakers sold, and R is the resulting sales revenue (in dollars), then apparently:

$$R = 75n$$

Once again, we observe that R is directly proportional to n, or R varies directly as n.

Now that we have seen some concrete examples direct variation, let's see how this type of relationship between two variables often leads us to consider problems involving proportions. This idea is contained in the next theorem.

Theorem-2: (Direct Variation)

Let y be **directly proportional** to x. If when x assumes the two different (non-zero) values x_1 and x_2, suppose y assumes the *corresponding* two values y_1 and y_2, then the following proportion is true:

$$\frac{y_2}{y_1} = \frac{x_2}{x_1} \tag{5}$$

That is, *if y is directly proportional to x, then the ratio of any two different values of x is equal to the ratio of the corresponding values of y.*

Example-6: (Hooke's Law)
According to Hooke's law, the distance that the spring stretches (away from its rest position) under the action of a force is *directly proportional* to the force applied to the spring. If a force of 5 pounds stretches a certain spring 3 inches, how far would the same spring stretch when a force of 14 pounds is applied?

Solution:
(a) Using the previous theorem, let x be the distance the spring stretches when 14 pounds of force are applied. Then:

$$\frac{x}{3} = \frac{14}{5}$$
$$5x = 3 \cdot 14 \quad \text{(Cross Multiply)}$$
$$5x = 42$$
$$\frac{5x}{5} = \frac{42}{5}$$
$$x = \frac{42}{5} \quad \text{Solution}$$

We conclude that under the influence of a force of 14 pounds, the same spring would stretch a distance $8\frac{2}{5}$ inches froms its rest position.

Example-7: The gross weekly wages earned by a part-time employee is directly proportional to the number of hours worked any week. If a part-time employee earned a gross pay of $220 after working 20 hours, then how much would the same employee earn after working 25 hours the following week?

Solution:
(a) Let x be the gross pay that the employee earns after working a total of 25 hours. Then by the previous theorem, we have:

$$\frac{x}{220} = \frac{25}{20}$$
$$20x = 220 \cdot 25 \quad \text{(Cross multiply)}$$
$$\frac{20x}{20} = \frac{220 \cdot 25}{20}$$
$$x = 11 \cdot 25$$
$$x = 275 \quad \text{Solution}$$

So after working a total of 25 hours, the same employee would earn a gross pay of $275 dollars.

A second way in which proportions naturally arise in various applications is when one variable is inversely related to another.

Definition-2: (Inverse Variation)

Given two variables x and y, we say that y is **inversely proportional** to x, or y **varies inversely** as x if and only if:

$$y = \frac{k}{x} \quad \text{(provided } x \neq 0\text{)} \qquad (6)$$

for some *non-zero* constant k.

Roughly speaking, two variables are said to be inversely related if they exhibit a "seesaw" relationship, i.e., as one variable goes up (or increases) the other variable goes down (or decreases). The next theorem tells us explicitly how inverse relationships lead to problems involving proportions:

Theorem-3: (Inverse Variation)

Let y be **inversely proportional** to x. If when x assumes the two different (non-zero) values x_1 and x_2, suppose y assumes the *corresponding* two values y_1 and y_2, then the following proportion is true:

$$\frac{y_2}{y_1} = \frac{x_1}{x_2} \qquad (7)$$

In other words, *if y is inversely proportional to x, then the ratio of any two different values of x is equal to the inverse ratio of the corresponding values of y.*

We end this lesson with an example of how Theorem-3 may be used to solve a typical problem which involves inverse variation.

Example-8: If a rectangle has a *constant* area, then its length y and width x are inversely proportional to each other. A given rectangle has a length of 18 and a width of 8. If its area remains fixed, but its width is increased to a new value of 9, what is the new length of the rectangle?

Solution:

(a) Let x be the new length of the rectangle. By the previous theorem, we have:

$$\frac{x}{18} = \frac{8}{9} \quad \text{(Theorem – 3)}$$

$$9x = 18 \cdot 8 \quad \text{(Cross multiply)}$$

$$\frac{9x}{9} = \frac{18 \cdot 8}{9}$$

$$x = 16 \quad \text{Solution.}$$

EXERCISE SET #8

PART A

1. A certain mathematics class contains 18 men and 20 women. Find the ratio of men to women.
2. The length of a given rectangle is 12 while its width is 10. Find the ratio of its length to width.
3. The population of a small country contains 540,000 women and 500,000 men. What is the ratio of women to men?
4. A dart player threw a dart at a target a total of 15 times. If she only hit the target 9 times, and missed it completely 6 times, what is the ratio of hits to misses?
5. Over a one year period, an investor earned an income of $1,200 on an initial investment of $3,600. Determine the income to investment ratio.
6. A small hotel has a total capacity of 120 rooms. If only a total of 55 rooms are occupied on a given date, then find the ratio of the number of rooms occupied to the number of rooms that are vacant.
7. A family restaurant enjoyed an annual gross income of $150,000. If its annual expenses totaled $35,000, what is its income to expense ratio?

In Exercises 8–15, solve for x in each proportion.

8. $$\frac{x}{4} = \frac{10}{5}$$

9. $$\frac{3}{4} = \frac{21}{x}$$

10. $$\frac{2x-1}{3} = \frac{10}{2}$$

11. $$\frac{x+1}{3} = \frac{14}{2}$$

12. $$\frac{3x+2}{5} = \frac{2x}{4}$$

13. $$\frac{4}{2(x-1)} = \frac{4}{7}$$

PART B

14. A 6-feet-tall man, standing in the sun, leaves a shadow 2 feet long. If the man is standing next to a building that has a shadow 27 feet long, how tall is the building?
15. A small scale model of an airplane is used to conduct wind tunnel tests. If the scale model has a wingspan of 15 inches, and is 3/70 the size of the actual plane, what is the wingspan of the actual plane?
16. A certain hotel consumed $375 worth of electricity during a given 3-day period. Assuming that it uses power at roughly a constant rate, how much (in dollars) electricity should it use in a 5 day period of time?
17. Show that if y is directly proportional to x, then x is also directly proportional to y.
18. Show that if y is inversely proportional to x, then x is also inversely proportional to y.
19. A given triangle has a base $b = 10$ and height $h = 8$. If a second triangle has a base $b = 20$, what must be the height of the second triangle so that it has the same area as the first?
20. If a body is initially at rest, and then experiences a constant accelerating force, its speed is directly proportional to the time during which the force is applied. After 2 seconds, a powerful engine causes a racing car to achieve a speed of 62 miles per hour. How fast would the race car go after a total of 5 seconds if its engine supplies the same constant force?
21. In a certain restaurant, the daily number of patrons who request any given meal is inversely proportional to its price. When the price of lobster was $16.00 per meal, a total of 10 customers requested it. If the price of lobster falls to $10.00 per meal, then how many customers will request it?
22. **Boyle's Law** states that when a gas is held at a constant temperature, then its volume V and pressure P are inversely proportional, i.e.,

$$PV = k$$

where k is a constant. A certain gas, that occupies a volume of 2 cubic feet at a pressure of 40 pounds per square inch, is compressed to a new volume of 0.3 cubic feet. Assuming that the temperature remains constant during the compression process, what is the pressure of the gas?
23. The daily revenue at a certain restaurant is approximately proportional to the number of customers it serves on a daily basis. If the restaurant earned $1,250 in daily revenue after serving a total of 125 customers, then about how much revenue should it receive after serving 150 customers?

24. Ken drove his automobile a total of 440 miles on 22 gallons of gasoline. If Ken were to go on a 700 mile trip, then about how many gallons of gasoline would the same car consume?
25. A certain windshield-washing solution is composed of one part alcohol and ten parts water by weight. How much of each liquid is needed to make two pounds of this solution?

9 Introduction to Word Problems

In this lesson, we will use what we have already learned about linear equations and proportions to solve some practical word problems. In order to solve such problems, the English sentences which appear in any given word problem must be eventually translated into mathematical language, and the mathematical statement of the problem is then solved using standard methods.

Although the methods used to solve word problems may vary, a general approach to solving most of the simple problems that we will consider is given below:

HOW TO SOLVE WORD PROBLEMS

Step #1: Read the problem carefully to determine what information is *given*, and determine what is the unknown quantity (or quantities) you are trying to find.

Step #2: Read the problem again.

Step #3: Write the unknown quantity(s) in terms of a single unknown variable x, and use the *given* information to write down an equation which involves the variable x.

Step #4: Solve the equation for x.

Step #5: Check your solution to see whether it is *reasonable*.

Step #6: After you have successfully solved the problem, *review the procedure* you used to solve it and try to remember your general approach for use on similar types of problems which you may encounter in the future.

Please don't underestimate the importance of Step #6. As you gain more experience, you will develop your own thought patterns that will help you to solve similar problems in the future.

Just like learning to play the piano, the more mathematics you do, the easier it is learn the next unfamiliar mathematical song. Well enough philosophy, let's look at how this procedure can be applied to some concrete examples.

Example-1: The sum of two numbers is 210. One number is twice the other number. Find each of the numbers.

Solution:
(a) The first two sentences here are the given facts about the problem: both numbers must add up to 210, and one number is twice the other number.
(b) Let x be the *smaller* of the two numbers. Since the other number is twice the first number, the second number must be $2x$; so we have:

$$x = \text{smallest number}$$
$$2x = \text{other number}$$

(c) Since the sum of both numbers is 210, we can write the equation:

$$x + 2x = 210$$
$$3x = 210$$
$$x = 70$$

(d) From (c), we conclude that one number is 70, while the other number is 2(70) or 140.

Example-2: A farmer decides to build a rectangular pasture with a total of 220 feet of fence. If the length of the pasture is to be 20 feet longer than its width, then what should be the dimensions of the pasture?

Solution:
(a) If we let x be the width of the pasture, then its length must be $x + 20$. The situation looks like the diagram below:

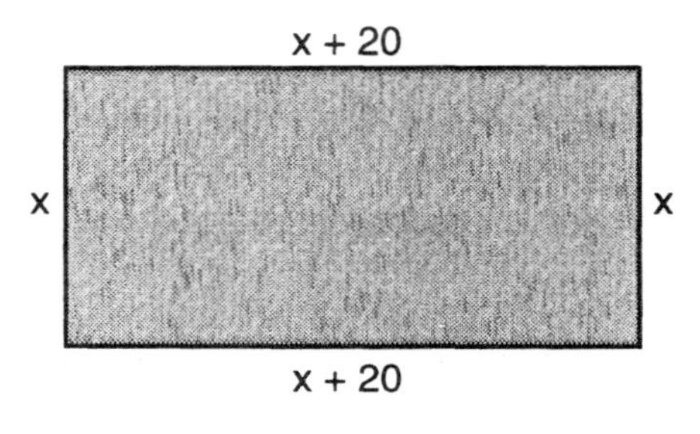

Example-2

(b) Now, the perimeter of the pasture is just the total distance around the rectangle. Since the farmer only has 220 feet of fence to work with, we have:

$$2(x + 20) + 2x = 220$$

and simplifing, we get:

$$4x + 40 = 220$$
$$4x = 180$$
$$x = \frac{180}{4}$$
$$x = 45$$

(c) We conclude that the width of the rectangle is $x = 45$ feet, while the length of the rectangle must be $x + 20$ or 65 feet.

Example-3: A student purchased a shirt at a local clothing store for a total cost of $44.94, an amount which included a 7% sales tax on the purchase. What was the cost of the shirt before tax?

Solution:
(a) The first sentence provides us with the given facts about this problem:

[Cost of Shirt (without tax)] + [Tax on Purchase] = $44.94

and, there is a 7% tax on the purchase.
(b) We're trying to find the cost of the shirt before taxes, so we will represent this unknown quantity by x:

Let x = Cost of Shirt (before tax)

(c) If we combine the information from (a) and (b), we can now write an equation which involves the unknown quantity x:

$$x + (.07)x = 44.94$$
$$(1.07)x = 44.94$$
$$x = \frac{44.94}{1.07}$$
$$x = 42.00$$

So we conclude that the cost of the shirt before tax is just $42.00.

Example-4: A man has a total of $1.05 of change in his pocket, in the form of nickels and dimes. If he has a total of 18 coins, then how many nickels and dimes does he have?

Solution:
(a) Here, the given information is apparent. If we let:

x = number of nickels
y = number of dimes

then, since the total amount of change is $1.05, we must have:

$$5x + 10y = 105 \qquad (1)$$

where we have written the equation in terms of cents to avoid decimals and to keep things as simple as possible.

(b) Now, Equation (1) is certainly correct, but it has one fatal flaw: we have *two* unknown quantities but only *one* equation! But we didn't yet use one crucial fact: the number of dimes plus the number of nickels must add up to 18. This means that we should be able to put everything in terms of just one unknown:

if x = number of nickels, then $(18 - x)$ = number of dimes

so Equation (1) may now be written in terms of just one unknown as:

$$5x + 10(18 - x) = 105$$
$$-5x + 180 = 105$$
$$-5x = -75$$
$$x = \frac{-75}{-5}$$
$$x = 15$$

(c) So we conclude that there must be $x = 15$ nickels, and $(18 - 15) = 3$ dimes. This checks out since:

$$15(5) + 3(10) = 105$$

Example-5: Two gunmen robbed the Second National Bank and fled the crime scene at 1:00 p.m., driving at a speed of 60 miles per hour. The police didn't get a description of the suspect's vehicle until later on, and they began pursuing the suspects at 2:00 p.m., at a speed of 80 miles per hour. At what time will the police overtake the suspects?

Solution:
(a) In this case, we see that the police will overtake the bank robbers when the distance travelled by the police car is equal to the total distance travelled by the bank robbers.
(b) If we let

x = total time travelled by bank robbers (in hours)

then, since the police didn't begin their pursuit until 2:00 p.m., we must have:

$(x - 1)$ = total time travelled by police (in hours)

(c) Since the total distance travelled is just the speed times the time, then we conclude that:

$60x$ = total distance travelled by bank robbers
$80(x - 1)$ = total distance travelled by police

(d) From our reasoning in (a) and (c), we see that the equation which must hold is:

distance travelled by police = distance traveled by robbers

$$\begin{aligned} 80(x-1) &= 60x \\ 80x - 80 &= 60x \\ 20x - 80 &= 0 \\ 20x &= 80 \\ x &= 4 \end{aligned}$$

Since $x = 4$ hours, and x represents the total time travelled by the robbers (who left at 1:00 p.m.) then the police will overtake the robbers at exactly 5:00 p.m..

Example-6: The sum of three consecutive even numbers is 36. What are the numbers?

Solution:
(a) In this problem, the given information is very simple, but we should be careful to notice the word "consecutive." By "consecutive" we mean "one right after the other".
(b) If we let:

$$x = \text{smallest even number}$$

then, then other two even numbers must be:

$$\begin{aligned} (x+2) &= \text{second even number} \\ (x+4) &= \text{third even number} \end{aligned}$$

(c) Since all three numbers must add up to 36, we get the equation:

$$\begin{aligned} x + (x+2) + (x+4) &= 36 \\ 3x + 6 &= 36 \\ 3x &= 30 \\ x &= 10 \end{aligned}$$

So the required numbers are: $x = 10$, $x + 2 = 12$, and $x + 4 = 14$.

EXERCISE SET #9

PART A

1. A certain mathematics class contains a total of 33 students. If there are 5 more men than women in the class, then how many men and women are there?
2. The sum of two numbers is 32. One number is three times the other number. What are the two numbers?
3. The sum of two consecutive odd numbers is 164. What are the numbers?
4. The sum of three consecutive even numbers is 66. What are the numbers?
5. A student has $1.80 as pocket change in the form of dimes and nickels. If the student has a total of just 20 coins, then how many dimes and nickels does she have?
6. In a certain recipe, a baker uses twice as much whole wheat flour as rye. If the total mixture contains 17 cups of flour, then how much of each type of flour was used?
7. Two trains leave their respective stations which are 360 miles apart, at 12:00 noontime. Accidently, both trains are on the same track travelling towards each other! If one train travels at 40 miles per hour, while the other train travels at 50 miles per hour, at what time will the trains collide?
8. A gardener decides to build a rectangular enclosure for his garden with a total of 16 feet of fence. If the length of the enclosure is 2 feet longer than its width, then what are the dimensions of the enclosure?
9. A student bought a CD at Jen's World of Music at a total cost of $17.28, an amount which included a 8% sales tax on the purchase. What was the cost of the CD before tax?
10. A mathematician is asked how old she is. She replies "my age 20 years ago will be exactly one-half of my age 5 years from today." How old is she today?

PART B

11. At 1:00 p.m., a train left Providence headed for Worcester at a speed of 30 miles per hour. At exactly 2:00 p.m., a second train left Worcester on a parallel track, headed for Providence at a speed of 30 miles per hour. If Worcester and Providence at 90 miles apart, then when will the trains meet?

12. Find four consecutive even numbers whose sum is 52.
13. A small boat can travel at a speed of 10 miles per hour in still water. To travel 45 miles downstream (moving with the current), the boat requires only 1/3 of the time it takes to travel the same distance moving upstream (traveling against the current). How fast is the river's current?
14. A student invests a part of $1,000 at 5%, and the remainder of the money at 4%. After one year, the income from the 4% investment exceeded the income from the 5% investment by a total of $22. Determine the amount invested at each rate.
15. A financial advisor earned a total of $2,650 for her client by purchasing a total of 450 shares of stock, spread between two different stocks: Stock-A and Stock-B. If the Stock-A earned $7.00 per share, while Stock-B earned $5.00 per share, then how many shares of each stock did she purchase?
16. An athlete can run at 6 miles per hour, and swim at a speed of 4 miles per hour. During his daily workout, it took him a total of 3 hours to travel a total distance of 16 miles. How long did he spend swimming? How long did he spend running?

10 Linear Inequalities

In the past few lessons, we have learned how to solve linear equations, and how to apply such equations to the solution of practical problems. In these lesson, we'll take a look at what are called linear inequalities, and find methods to solve them.

Before proceeding officially with inequalities, we should probably review the inequality symbols that are conventionally used in mathematics:

Table-1: Inequality Symbols	
Inequality Symbol	**What the Symbol Means**
$a < b$	a is less than b
$a \leq b$	a is less than or equal to b
$a \neq b$	a is not equal to b
$a > b$	a is greater than b
$a \geq b$	a is greater than or equal to b

An easy way of remembering what the statements $a < b$ and $a > b$ mean is noticing that in each case, *the tip of the inequality symbol always points towards the smaller of the two numbers.*

Consequently, if we are given a statement like $a > b$, then since the tip of the inequality symbol points towards b, we conclude that a must be greater than b. In order to get some practice interpreting these symbols, try working out the next example.

Example-1: Determine whether each of the following statements is true or false.

(a) $2 + 2 > 5$
(b) $2 \leq 2$
(c) $4 + 1 \geq 4$
(d) $4 + 3 \neq 4$

Solution:

(a) False, since $2 + 2 = 4$ and 4 is not greater than 5.
(b) This statement is true since 2 is less than **or** equal to 2.
(c) True, since $4 + 1 = 5$ and $5 \geq 4$.
(d) True, since $4 + 3 = 7$ and $7 \neq 4$.

An **inequality** is nothing more than any two expressions that are joined by one of the inequality symbols in the above table. Thus, the following statements are examples of inequalities:

$$2 + 1 < 4$$
$$2x + 1 > 5$$

Notice that that first inequality is always true, while the second inequality is only true when $x > 2$. For this reason, we call the second statement a **conditional inequality** since it is only true under certain conditions, i.e., for certain values of the variable x.

Loosely speaking, a **linear inequality** is a conditional inequality where the unknown variable x only appears to the *first power.* A formal definition of this concept is given below:

Definition-1: (Linear Inequality in One Variable)

A **linear inequality** (in one variable) is a conditional inequality which can be written in the form:

$$ax + b < cx + d \tag{1}$$

or any inequality which can be obtained by replacing the "$<$" symbol in (1) with any other inequality symbol in Table-1. For our purposes, we agree that a,b,c, and d are real numbers and not both a and c are zero.

Example-2: Determine whether the following inequalities are linear inequalities:

(a) $2x - 5 < 4$
(b) $3x - 1 \geq 0$
(c) $x^2 + x < 2x$

Solution:

(a) This is a linear inequality since if we take:

$$a = 2, b = -5, c = 0, d = 4$$

in (1) we get:

$$(2)x + (-5) < (0)x + (4) \text{ or } 2x - 5 < 4$$

(b) This is also a linear inequality since any inequality symbol from Table-1 can be used to connect the lefthand and righthand sides of (1). Here, if we take:

$$a = 3, b = -1, c = 0, d = 0$$

in (1), then we get:

$$3x + (-1) \geq (0)x + (0) \text{ or } 3x - 1 \geq 0$$

(c) This is not a linear inequality due to the presence of the x^2 term. Notice that in (1), x only appears to the first power.

A *solution* to a linear inequality is any value of x which when substituted into the inequality, produces a true statement. The **solution set** S of a linear inequality is the set of *all* possible values of x for which the inequality holds true.

For example, consider the linear inequality given by:

$$2x \geq 2$$

By direct substitution, it is easy to see that $x = 1$ and $x = 2$ are solutions of this inequality since:

$$2(1) \geq 2 \text{ and } 2(2) \geq 2$$

are both true statements. Since, it is also clear that *any* real number $x \geq 1$ is a solution of this inequality; we conclude that the solution set S for this inequality is given by:

$$S = \{x \mid x \in R \text{ and } x \geq 1\}$$

Although technically, we should write our solution sets using set notation, for the sake of brevity, we shall agree to simply write the solution as $x \geq 1$.

In this simple example, observe that the solution set S is an infinite set. This means that, unlike linear equations, linear inequalities may have *infinitely many* solutions. As we shall see in what follows, however, the method of obtaining the solution set of an inequality is quite similar to the process involved in solving a linear equation. Take a look at the following theorem:

Theorem-1: (Addition/Subtraction Property)

If the inequality:

$$a < b$$

holds true, then the following inequalities are also true for any real number c:

$$a + c < b + c \qquad (2)$$

$$a - c < b - c \qquad (3)$$

In other words, *the same number may be added (or subtracted) from both sides of a true inequality, and the result is a true inequality.*

Example-3: Solve the following inequalities:
(a) $x + 5 < 9$
(b) $x - 10 \geq 1$

Solution:
(a) Using Equation (3) in Theorem-1, we get:

$x + 5 < 9$	
$x + 5 - 5 < 9 - 5$	Subtract 5 from both sides.
$x < 4$	Solution

(b) Using Equation (2) in Theorem-1, we obtain:

$x - 10 \geq 1$	
$x - 10 + 10 \geq 1 + 10$	Add 10 to both sides.
$x \geq 11$	Solution

Theorem-2: (Multiplication/Division Property #1)

If the inequality:

$$a < b$$

is true, then for any *positive* real number c, then both of the following inequalities are also true:

$$a \cdot c < b \cdot c \quad (4)$$

and,

$$\frac{a}{c} < \frac{b}{c} \quad (5)$$

In other words, *if both sides of a true inequality are either multiplied or divided by a positive number, then a true inequality is obtained.*

Example-4: Solve the following inequalities:
(a) $3x + 1 < x - 4$
(b) $5x - 7 \geq 2x + 3$

Solution:
(a) We have:

$3x+1<x-4$	
$3x+1-1<x-4-1$	Subtract 1 from both sides.
$3x<x-5$	Simplify
$3x-x<x-5-x$	Subtract x from both sides.
$2x<-5$	Simplify
$\frac{2x}{2}<\frac{-5}{2}$	Divide both sides by 2.
$x<-\frac{5}{2}$	Solution.

(b) Similarly, we obtain:

$5x-7\geq 2x+3$	
$5x-7+7\geq 2x+3+7$	Add 7 to both sides.
$5x\geq 2x+10$	Simplify.
$5x-2x\geq 2x+10-2x$	Subtract $2x$ from both sides.
$3x\geq 10$	Simplify.
$\frac{3x}{3}\geq\frac{10}{3}$	Divide both sides by 3.
$x\geq\frac{10}{3}$	Solution.

So far, we see that solving linear inequalities is just like solving linear equations, but we have to be careful. Their is one important exception to this observation and it is given in the next theorem:

Theorem-3: (Multiplication/Division Property #2)

If the inequality:

$$a < b$$

is true, then for any *negative* number c, then the following inequalities are also true:

$$a \cdot c > b \cdot c \tag{6}$$

and,

$$\frac{a}{c} > \frac{b}{c} \tag{7}$$

That is, *if both sides of a true inequality are either multiplied (or divided) by a* **negative** *number, then a true inequality is obtained; however, the inequality sign in the new inequality must be* **reversed.**

Example-5: Solve the following inequalities:
(a) $-5x + 1 < 2x + 7$
(b) $3x - 2 \geq 6x + 3$

Solution:
(a) We obtain:

$-5x+1<2x+7$	
$-5x+1-1<2x+7-1$	Subtract 1 from both sides.
$-5x<2x+6$	Simplify.
$-5x-2x<2x+6-2x$	Subtract $2x$ from both sides.
$-7x<6$	Simplify.
$\frac{-7x}{-7}>\frac{6}{-7}$	Divide both sides by -7 and reverse the direction of the inequality symbol.
$x>-\frac{6}{7}$	Solution.

(b) Similarly:

$3x-2\geq 6x+3$	
$3x-2+2\geq 6x+3+2$	Add 2 to both sides.
$3x\geq 6x+5$	Simplify.
$3x-6x\geq 6x+5-6x$	Subtract $6x$ from both sides.
$-3x\geq 5$	Simplify.
$\frac{-3x}{-3}\leq \frac{5}{-3}$	Divide both sides by -3 and reverse the direction of the inequality symbol.
$x\leq -\frac{5}{3}$	Solution.

EXERCISE SET #10

PART A

In Exercises 1–10, solve each inequality.

1. $2x + 1 < 5$
2. $3x - 2 \leq 7$
3. $-7x > 49$
4. $-\frac{2}{3}x + 2 \leq \frac{1}{6}$
5. $2x + 5 \geq 7x + 3$
6. $\frac{x-4}{5} \geq \frac{2x-2}{3}$
7. $1 + \frac{x}{2} < \frac{3x+1}{4}$
8. $\frac{2}{3}x \leq 4$
9. $2(x+3) - x > -3x + 4$
10. $5(x+2) \leq 10(x+1)$
11. A student received the grades of 70 and 80 on his two mathematics quizzes. What score must he obtain on the third quiz if he is to maintain a final average which is greater than 80?
12. Does the inequality given by:

    $$2x + 1 < 2x - 1$$

    have any solution? Explain your answer.
13. A certain company manufactures key chains that sell for $2.00 each. This means that if x key chains are sold, then the revenue R (in dollars) generated is:

    $$R = 2x$$

    The cost C (in dollars) of manufacturing x keychains is given by:

    $$C = \frac{1}{2}x + 300$$

    In order to realize a profit, the net revenue ($R - C$) must be a positive number. How many keychains must be sold if the company is to make a profit?
14. If a baseball is thrown vertically upward, its speed v (feet per second) at any time t (sec.) is given by:

    $$v = v_0 - 16t$$

where v_0 (feet per second) is its upward speed at the instant it is released. If a given baseball has an initial upward speed of $v_0 = 64$ feet per second, then for what values of t will its speed v be positive?

15. In order to pass a safety inspection, a certain small truck cannot carry a total weight (cargo plus driver) greater than 2400 pounds. A driver who weighs 200 pounds wishes to load the truck with boxes of coffee that weigh 50 pounds each. What is the greatest amount of coffee (in boxes) which can be loaded without exceeding the safety standard?

PART B

16. Prove (5) in Theorem-2.
17. Prove (7) in Theorem-3.
18. Find all values of x which satisfy the inequality:

$$\frac{1}{5} \leq \frac{2x-1}{2} \leq \frac{3}{10}$$

[**Hint:** Work on all three parts of the inequality at the same time.]

19. Find the values of x which satisfy the inequality:

$$|x - 1| < 2$$

[**Hint:** Use the definition of absolute value, and break up the problem into two separate cases according to whether $(x - 1) \geq 0$ or $(x - 1) < 0$.]

20. Find the values of x which satisfy the *non-linear* inequality:

$$x^2 - 5x + 6 > 0$$

by using the relation:

$$x^2 - 5x + 6 = (x - 2)(x - 3)$$

and the fact that the product of two numbers is positive if and only if both numbers have the same sign.

21. If a and b are real numbers with $b \geq 0$, find the solution of the inequality:

$$|x - a| < b$$

22. Find the solution of the inequality:

$$\frac{|x-1|}{|x+1|} < 1$$

[**Hint:** On the real number line, $|x - 1|$ is just the distance between x and 1, while $|x + 1|$ is the distance between x and -1.]

11 INTRODUCTION TO FUNCTIONS

Like the concept of a set, the idea of a function is central to all of mathematics. A function may be viewed as a special type of relationship between two (or more) variables. Let's begin with a simple definition.

Definition-1 (Function)

Let two variables x and y be related in such a manner that for each value of x there corresponds exactly *one* value of y, then we say that y is a **function** of x and write:

$$y = f(x) \text{ (which is read "y is a function of x")} \qquad (1)$$

Furthermore, we call x the **independent variable,** and y the **dependent variable** in (1).

The notation $y = f(x)$ does not mean that f is being multiplied by x; it is simply shorthand notation which tells us that y is a function of x. We could also use the symbols:

$$y = g(x) \text{ and } y = h(x) \text{ etc.}$$

to accomplish the same thing, i.e., the letters f, g, h merely serve to name different functions of x just like we might use the different letters A,B,C to represent various sets.

We shall agree that we will only deal with **real-valued** functions, i.e., if we are given a function $y = f(x)$, then we shall assume that the values for x and the corresponding values for y can only be real numbers.

Example-1: If a baseball is released from rest, then the distance s (feet) that it falls after t (seconds) is given by:

$$s = 16t^2$$

Is s a function of t?

Solution:
(a) We choose some arbitrary values for t, and then calculate the corresponding values of s to obtain Table-1 below:

TABLE-1: THE FALLING BASEBALL

Values of t	Corresponding Values of s
1	$16(1)^2 = 16$
2	$16(2)^2 = 64$
3	$16(3)^2 = 144$
4	$16(4)^2 = 256$
5	$16(5)^2 = 400$

(b) It is obvious that we could continue in this fashion indefinitely, and select even more values for t and then calculate the corresponding values for s. But one thing is very clear from looking at Table-1: for each value of t we only get exactly *one value* of s. Thus, based upon Definition-1 above, we must conclude that s is a function of t, and write:

$$s = f(t) = 16t^2$$

In this example, t is the independent variable, and s is the dependent variable since its value depends upon t.

Example-2: The relationship between temperature in degrees Fahrenheit F and degrees Celsius C is given by:

$$F = \frac{9}{5}C + 32$$

Is F a function of C?

Solution:
(a) As before, we choose some arbitrary values for C and calculate the corresponding values of F:

TABLE-2: RELATION BETWEEN FAHRENHEIT TEMPERATURE AND CELSIUS TEMPERATURE	
Centigrade Temp. *C*	**Corresponding Fahrenheit Temp. *F***
0	$\frac{9}{5}(0)+32=32$
20	$\frac{9}{5}(20)+32=68$
40	$\frac{9}{5}(40)+32=104$
60	$\frac{9}{5}(60)+32=140$
80	$\frac{9}{5}(80)+32=176$

(b) From the table it is apparent that for each value of C there will correspond exactly one value of F so that F is a function of C and we can write:

$$F=f(C)=\frac{9}{5}C+32$$

Here, C is the independent variable, and F is the dependent variable.

Example-3: Given that x and y are related according to:

$$y^2 = x$$

Determine whether y is a function of x.

Solution:
(a) As before, we construct a table to see how many values of y we get for each value of x:

TABLE-3: THE RELATION: $Y^2 = X$	
Values of x	**Corresponding Values of y**
0	0
1	−1
1	1
4	−2
4	2

(b) In this case, we notice that for some values of x we get more than one value for y. For example, if $x = 1$ then we have both $y = -1$ and $y = 1$. We conclude that y is *not* a function of x.

From the previous examples, we see that every function is a **relation,** i.e., but it is a special type of relation between two variables x and y such that for each value of x there corresponds a unique (one and only one) value of y. On the other hand, from Example-3, we see that *not all relations are functions.*

Definition-2: (Functional Notation)

Let y be a function of x, or symbolically: $y = f(x)$. Suppose that when x assumes some value $x = a$, the corresponding value of y is $y = b$, then we agree to use shorthand **functional notation** to write:

$$b = f(a) \tag{2}$$

which is read "f at a equals b."

Example-4: Given the function:

$$f(x) = x^2 + 1$$

calculate each of the following:

(a) $f(0)$
(b) $f(2)$
(c) $f(-1)$

Solution:
(a) Using Definition-2, $f(0)$ is the value of y we obtain when we substitute $x = 0$ into the above equation:

$$f(0) = (0)^2 + 1$$
$$f(0) = 1$$

(b) Similarly,

$$f(2) = (2)^2 + 1$$
$$f(2) = 5$$

(c) And,

$$f(-1) = (-1)^2 + 1$$
$$f(-1) = 2$$

Example-5: Given that:

$$y = g(x) = \sqrt{x+1}$$

calculate:

(a) $g(0)$
(b) $g(3)$
(c) $g(24)$

Solution: Proceeding as in the previous example, we obtain:

(a) $$g(0) = \sqrt{0+1} = \sqrt{1} = 1$$

(b) $$g(3) = \sqrt{3+1} = \sqrt{4} = 2$$

(c) $$g(24) = \sqrt{24+1} = \sqrt{25} = 5$$

Definition-3: (Domain and Range)

Let y be a function of x, or $y = f(x)$. Then the **natural domain** or **domain** of the function is the set of all possible values of x for which y is *defined*. We shall agree to denote the domain of f by the symbol:

$$Domain(f) \quad \text{(which is read the "Domain of f")} \tag{3}$$

The **range** of f, denoted by *Range(f)*, is the set of all corresponding values of y as x assumes its values in the domain. Formally,

$$Range(f) = \{y \mid y = f(x) \text{ where } x \in Domain(f)\} \tag{4}$$

Example-6: Find the domain and range of the function:

$$y = f(x) = \frac{1}{x}$$

Solution:
(a) It is clear that the domain cannot include the value $x = 0$ since division by zero is undefined; we conclude that:

$$Domain(f) = R - \{0\}$$

That is, the domain consists of all real numbers $x \neq 0$.
(b) To find the range, notice that y can never be zero since the numerator is a constant. But it is also clear that y can assume all negative and positive values at will; thus,

$$Range(f) = R - \{0\}$$

In other words, the range consists of all non-zero real numbers as well.

Example-7: Find the domain and range of the function:

$$y = g(x) = x^2 + 1$$

Solution:
(a) Here, the function is defined for all possible real values of x, so we conclude that:

$$Domain(g) = R$$

or, the domain consists of all real numbers.
(b) Since the square of any real number is always non-negative, we see that the smallest possible value of y is one; consequently, the is range is given by:

$$Range(g) = \{y \mid y \in R \text{ and } y \geq 1\}$$

or, stated differently, the range consists of all real numbers that are greater than or equal to one.

Example-8: Find the domain and range of the function:

$$y = h(x) = \sqrt{x - 1}$$

Solution:
(a) Since this function involves a square root we have to be careful. **Observe that the square root of a negative number cannot be a real number – since any real number squared is always non-negative.** Thus, we must exclude all values of x which might make the radicand negative; the domain is therefore:

$$Domain(h) = \{ x \mid x \in R \text{ and } x \geq 1\}$$

or all real numbers $x \geq 1$.
(b) Since the square root of a real number is assumed to be the principal square root (which is non-negative), then the range of this function is clearly:

$$Range(h) = \{ y \mid y \in R \text{ and } y \geq 0\}$$

or all real numbers $y \geq 0$.

EXERCISE SET #11

PART A

In Exercises 1–4, determine whether y is in function of x:

1. $y = 3x + 5$
2. $y^4 = x$
3. $y = x^2 + x + 1$
4. $y = \sqrt{x}$
5. Given the function:

$$y = g(x) = x^3$$

calculate the following:

(a) $g(0)$
(b) $g(-1)$
(c) $g(2)$
(d) $g(1)$

6. Given the function:

$$y = f(x) = \frac{x+1}{x-1}$$

calculate the following:

(a) $f(0)$
(b) $f(2)$
(c) $f(-1)$
(d) $f(3)$

In Exercises 7–12, find the domain and range of each of the following functions:

7. $y = f(x) = \sqrt{2x+1}$
8. $y = g(x) = 1/(x + 2)$
9. $y = f(x) = x^2 - 2$
10. $y = h(x) = 1/\sqrt{x-2}$
11. $y = g(x) = 1$
12. $y = h(x) = 1/(x^2 + 1)$

PART B

13. From a previous example, we know that the relationship between temperature in degrees Fahrenheit F and degrees Celsius C is given by:

$$F = \frac{9}{5}C + 32$$

(a) Solve this equation for C thereby expressing the Celsius temperature C as a function of F.
(b) Based on your answer in (a), determine what Celsius temperature corresponds to 212 degrees Fahrenheit.

14. From Example-1, we know that if an object is released from rest, then the distance s (feet) that it falls after t (seconds) is given by:

$$s = 16t^2$$

A student dropped a baseball from the top of the Empire State Building and timed its fall. If it took the baseball a total of 9.5 seconds to hit the pavement, then how tall is this building?

15. If an object is released from rest, then its speed v (feet per second) after t seconds is given by the function:

$$v = g(t) = 32t$$

(a) Using the information from Exercise-14, determine how fast the baseball was traveling (in feet per second) when it struck the pavement.
(b) Using the fact that:

$$60\frac{\text{miles}}{\text{hour}} = 88\frac{\text{feet}}{\text{second}}$$

convert your answer in (a) to miles per hour. This should convince you that it's not a good idea to drop things off the top of the Empire State Building!

16. In a small manufacturing firm, the cost of manufacturing a certain type of chair is $20 per chair, along with a fixed (one-time) cost of $1,200 for machine setup.

(a) Express the total cost $C(x)$ of producing x chairs as a function of x. [We call $C(x)$ a **cost function.**]
(b) What is the cost of producing 100 chairs?

17. The number of bushels of corn demanded at a price of p dollars per bushel is given by the function:

$$D(p) = \frac{3000}{p} \quad (\text{where } p > 0)$$

In economics, we call $D(p)$ a **demand function.**

(a) Calculate $D(30)$.
(b) What is the meaning of your answer in (a)?

18. A function $y = f(x)$ is said to be **one-to-one** if distinct values of x always give rise to distinct values of y. That is, $y = f(x)$ is one-to-one if and only if:

$$x_1 \neq x_2 \text{ implies that } f(x_1) \neq f(x_2)$$

for all $x_1, x_2 \in Domain(f)$. Show that the function:

$$y = f(x) = ax + b \quad (\text{where } a \neq 0)$$

where a and b are constants, is a one-to-one function.

19. A function $y = f(x)$ is said to be **strictly increasing** if increasing values of x produce increasing values of y. More precisely, we say that $y = f(x)$ is strictly increasing if and only if:

$$x_1 < x_2 \text{ implies that } f(x_1) < f(x_2)$$

for all $x_1, x_2 \in Domain(f)$. Show that the function:

$$y = f(x) = ax + b \quad (\text{where } a > 0)$$

where a and b are constants, is strictly increasing.

20. A automobile travels at a constant speed of 65 miles per hour. Express the distance d (miles) it travels as a function of its travel-time t (hours).
21. A rectangular pasture has a fixed length of 400 feet. Express the area A of the pasture as a function of its width w (feet).
22. We say that z is a **multivariate function** of x and y if for each ordered pair (x,y) of values for x and y, there corresponds one and only one value of z. We can generalize our functional notation and write this special relationship as: $z = f(x,y)$. Given that:

$$z = f(x,y) = x^2 + xy + 1$$

Calculate each of the following:

(a) $f(1,2)$
(b) $f(-1,1)$
(c) $f(3,2)$

23. If we ignore frictional forces, it is known that when a roller-coaster car, which is initially at rest, falls from a height of h (feet) it achieves a ground speed v (feet per second) of:

$$v = f(h) = \sqrt{64h}$$

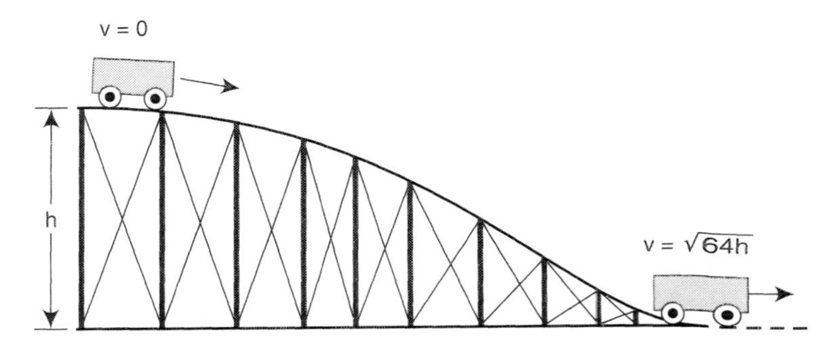

Exercise-23: A Monster Rollercoaster

In a "monster rollercoaster", a rollercoaster car slides down a height of 144 feet. What is its ground speed at the very bottom?

24. An **infinite sequence** is just a special type of function whose domain is the set of Natural Numbers N. As the independent variable n assumes the successive values $n = 1,2,3,\ldots$, the corresponding values of the function are listed as an ordered set of numbers. A very famous sequence is called the **Fibonacci sequence** which consists of the infinite ordered set of **Fibonacci numbers:**

THE FIBONACCI SEQUENCE

$$1,1,2,3,5,8,13,21,\ldots$$

The n-th Fibonacci number $f(n)$ can be obtained from the formula:

$$f(1) = 1,\ f(2) = 1$$
$$f(n + 2) = f(n) + f(n + 1)$$

(a) Use this formula to find the next three Fibonacci numbers.
(b) By examining the ratios of larger and larger successive Fibonacci numbers:

$$\frac{1}{1} = 1.0,\quad \frac{2}{1} = 2.0,\quad \frac{3}{2} = 1.5$$
$$\frac{5}{3} = 1.66\ldots,\quad \frac{8}{5} = 1.6,\quad \frac{13}{8} = 1.625$$
$$\frac{21}{13} = 1.6153846,\quad \text{etc.}$$

show that these ratios get closer and closer to the number:

$$\frac{1+\sqrt{5}}{2} = 1.618034\ldots$$

which is called the **Golden Ratio.** This special number has played a prominent role in architecture and art for over 2,000 years.

12 Rectangular Coordinates

Before we continue our study of functions, we must take a short detour in our journey to examine the rectangular coordinate system. As you will see in the next lesson and others that follow, this system will help us to graphically visualize how certain functions behave.

The **rectangular (or Cartesian) coordinate system** was invented by the French mathematician Rene Descartes (1596–1650) and it provides us with a means of locating any mathematical point in the plane (a flat surface) with two numbers, called the **coordinates** of the given point.

The rectangular coordinate system is made up of two axes: a horizontal **x-axis,** and a vertical **y-axis.** These axes intersect at a single point which we call the **origin O** of the coordinate system. This state of affairs is depicted below in Figure-1.

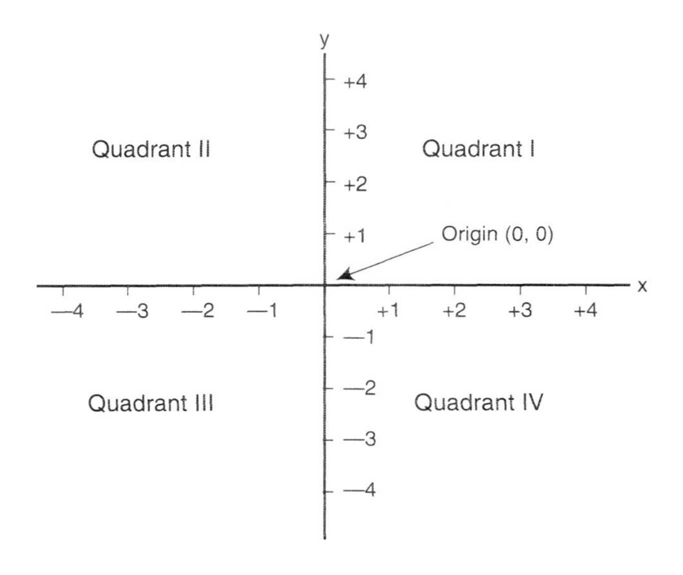

Figure 1. Rectangular Coordinate System

More precisely, the position of a given point in the plane is completely defined by an **ordered pair** of numbers (x,y). We use the term *ordered* pair because the order of the numbers is very important. The first number in the ordered pair represents the x-coordinate of a given point, while the second number gives the y-coordinate of that point. In order to plot any point in the xy-plane, we follow this simple procedure:

HOW TO PLOT ANY POINT (X,Y) IN THE PLANE

Step #1: We begin at the origin $O(0, 0)$ of the rectangular coordinate system.

Step #2: If the x-coordinate is a positive number, then move to the right x units. If the x-coordinate is a negative number, then move to the left $|x|$ units. Finally, if the x-coordinate is zero, then stay at the origin.

Step #3: If the y-coordinate is a positive number, then move upward y units. If the y-coordinate is a negative number, then move downward uyu units. If the y-coordinate is zero, then stay put; in either case, you've now found the point.

Example-1: Plot the point $P(2, 3)$.

Solution:

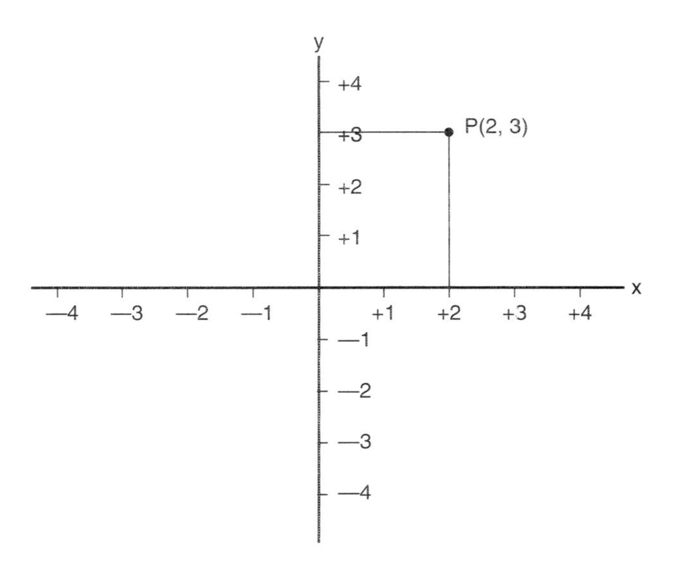

Figure-2: Plotting the Point P(2,3)

In order to plot the point $P(2, 3)$, we first start at the origin. Since $x = 2$ is positive, we move 2 units to the right. Since the y-coordinate is also positive, we finally move 3 units upward. This point lies in the first quadrant of the Cartesian coordinate system.

Example-2: On the same diagram, plot the additional points:
(a) $Q(-4,4)$
(b) $R(-3,-1)$
(c) $T(3,-3)$

Solution:

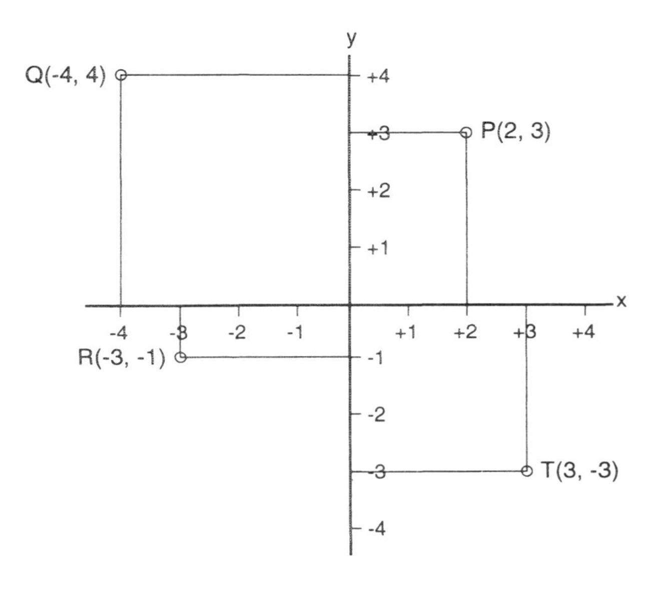

Figure-3: The Solution to Example-2

(a) In order to plot the point $Q(-4,4)$, we first start at the origin. Since $x = -4$ is negative, we move $|-4| = 4$ units to the left. Since the y-coordinate is positive, we finally move 4 units upward. Observe that this point lies in the second quadrant of our coordinate system.
(b) Similarly, to plot the point $R(-3,-1)$, we again start at the origin. Since $x = -3$ is negative, we move $|-3|$ or 3 units to the left. Since the y-coordinate is negative, we now move $|-1| = 1$ unit downward. Observe that this point lies in the third quadrant of the Cartesian coordinate system.
(c) Finally, in order to plot the point $T(3,-3)$, we start at the origin. Since $x = 3$ is positive, we move 3 units to the right. Since the y-coordinate is negative, however, we now move $|-3| = 3$ units downward. Notice that this point lies in the fourth quadrant of our coordinate system.

In addition to just locating various points in the plane, we can do many interesting things with the Cartesian coordinate system. One of these things is finding the distance between two given points in the plane. But before we do this, let's review the **Pythagorean Theorem.**

The Pythagorean Theorem is an extremely important statement which tells us the relationship between the lengths of the sides of a right triangle. A right triangle is just a triangle that contains a right (or 90°) angle. Take a look at Figure-4 below.

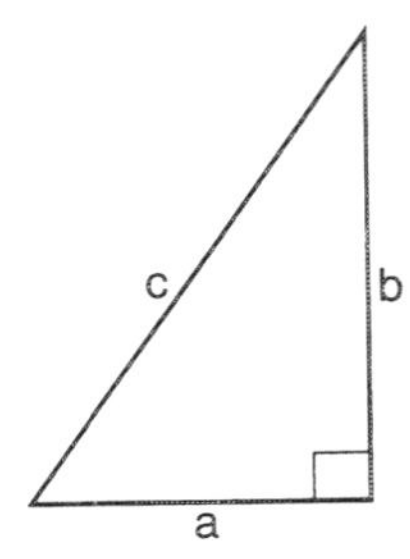

Figure-4: A Typical Right Triangle

We call the side (of length) c the **hypotenuse,** and this is always the side which is directly *opposite* the right angle. The other two sides (of length) a, and b respectively, are called **legs** of the right triangle. We now state the Pythagorean Theorem:

Theorem-1: (Pythagoras)

In any right triangle which has a hypotenuse of length c, and legs of lengths a and b respectively, we have the relationship:

$$c^2 = a^2 + b^2 \tag{1}$$

or equivalently,

$$c = \sqrt{a^2 + b^2} \tag{2}$$

Let's see how this theorem works in a simple example.

Example-3: A right triangle has legs of lengths $a = 2$, and $b = 3$ respectively. What is the length of its hypotenuse?

Solution: Here, we merely substitute the values of $a = 2$ and $b = 3$ into Equation (2) above to obtain:

$$c = \sqrt{2^2 + 3^2} = \sqrt{4 + 9} = \sqrt{13}$$

We conclude that the hypotenuse has a length of $c = \sqrt{13}$ units, or $c \approx 3.61$, where the symbol "$\approx$" means "is approximately equal to."

With the Pythagorean Theorem safely behind us, we can now derive a formula for calculating the *distance* between any two points in the plane.

Theorem-2: (Distance Formula)

Let $P_1(x_1,y_1)$ and $P_2(x_2,y_2)$ be any two points in the plane, then the distance d between these points is given by:

$$d = \sqrt{(x_2 - x_1)^2 + (y_2 - y_1)^2} \qquad (3)$$

We now complete this lesson with a final example.

Example-4: Find the distance between each pair of points:
(a) $A(3,2)$ and $B(6,4)$
(b) $P(1,-1)$ and $Q(3,0)$
(c) $C(0,-2)$ and $D(3,2)$

Solution:
(a) It doesn't matter which point we choose to label as "Point-1" or "Point-2" as long as we stick to our choice once we've made it. In this case, we will take A as the first point, and B as the second point; consequently,

Point #1 : $A(3,2)$ so that we let: $x_1 = 3$, and $y_1 = 2$
Point #2 : $B(6,4)$ so that we let: $x_2 = 6$, and $y_2 = 4$

Now, using the distance formula, we obtain:

$$d = \sqrt{(x_2 - x_1)^2 + (y_2 - y_1)^2} = \sqrt{(6-3)^2 + (4-2)^2}$$
$$= \sqrt{3^2 + 2^2} = \sqrt{13}$$

(b) Similarly, for the points $P(1,-1)$ and $Q(3,0)$ we obtain:

$$d = \sqrt{(x_2 - x_1)^2 + (y_2 - y_1)^2} = \sqrt{(3-1)^2 + (0-(-1))^2}$$
$$= \sqrt{2^2 + 1^2} = \sqrt{5}$$

(c) Finally, for the points $C(0,-2)$ and $D(3,2)$, we obtain:

$$d = \sqrt{(x_2 - x_1)^2 + (y_2 - y_1)^2} = \sqrt{(3-0)^2 + (2-(-2))^2}$$
$$= \sqrt{3^2 + 4^2} = \sqrt{25} = 5$$

EXERCISE SET #12

PART A

In Exercises 1–4, plot each point in the plane.

1. $A(1,3)$
2. $B(2,-3)$
3. $C(-3,2)$
4. $D(3,-4)$

In Exercises 5–10, calculate the length of the hypotenuse of a right triangle whose legs have the given values:

5. $a = 3, b = 4$
6. $a = 6, b = 8$
7. $a = 2, b = 1$
8. $a = 3, b = 5$
9. $a = 4, b = 6$
10. $a = 7, b = 3$

In Exercises 11–15, calculate the distance between each pair of points.

11. $A(1,1)$ and $B(2,2)$
12. $C(-1,2)$ and $D(3,4)$
13. $E(-3,0)$ and $F(0,4)$
14. $P(2,6)$ and $Q(1,-1)$
15. $R(-2,-2)$ and $S(3,3)$

PART B

16. On a certain map, the rectangular "map coordinates" of cities A and B are (1,3) and (7,11) respectively. If one unit of distance on the map is equal to 10 miles, then how far apart are these cities?
17. A "Cartesian bug" starting at the origin, crawls 12 steps to the right (along the positive x-axis), and then 16 steps downward (in the negative y-direction). How far away is the bug from the origin?
18. Show that the distance d between two points $P_1(x_1,y_1)$ and $P_2(x_2,y_1)$ which have the *same* y-coordinates is given by:

$$d = |x_2 - x_1|$$

Hint: You will need the fact that for any real number a, we can write:

$$|a| = \sqrt{a^2}$$

19. Show that the distance d between two points $P_1(x_1,y_1)$ and $P_2(x_1,y_2)$ which have the *same* x-coordinates is given by:

$$d = |y_2 - y_1|$$

20. Show that the coordinates of the **midpoint** $M(x_m, y_m)$ of the line segment whose endpoints are $P_1(x_1, y_1)$ and $P_2(x_2, y_2)$ is given by the formula:

$$x_m = \frac{x_1 + x_2}{2} \text{ and } y_m = \frac{y_1 + y_2}{2} \quad (4)$$

Hint: Use Equation (3) to show that the distance between P_1 and M is exactly *one-half* the distance between P_1 and P_2.

21. Use Equation (4) of Exercise-20 to find the coordinates of the midpoint of the line segment whose endpoints are $A(2,3)$ and $B(10,7)$.
22. Prove that for a right triangle, the midpoint of its hypotenuse is equidistant (or the same distance away) from each of its three vertices. **Hint:** To make things easy, take the vertices as $A(0,0)$, $B(a,0)$ and $C(a,b)$; and draw a picture.
23. Starting at the point $P_1(1,2)$, a "Cartesian Bug" crawls in a straight line to another point $P_2(13,18)$. If it takes the bug 5 seconds to complete his journey, and if we use the "foot" as our unit of length, then what is the bug's speed?
24. A triangle is inscribed in a semicircle as shown in the figure below:

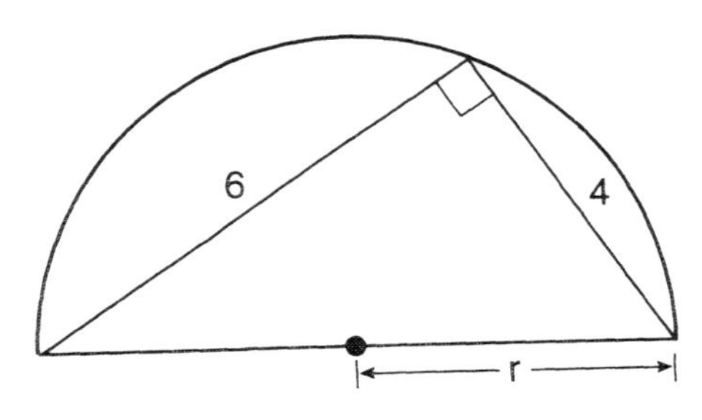

Exercise-24

From geometry, it known that whenever a triangle is inscribed in a semicircle as shown above, then it will always be a right triangle. Use this fact to find the radius r of the semicircle.

25. It may be shown that the *converse* of the Pythagorean Theorem holds true, i.e., if a given triangle has sides $a,b,$ and c, and if these sides are related by the formula:

$$c^2 = a^2 + b^2$$

then the given triangle must be a *right* triangle. A given triangle has sides of length 30, 24, and 18. Is it a right triangle?

13 Graphing Linear Functions

In this lesson, we'll learn how to graph *linear functions*. Loosely speaking, a linear function is a special function such as:

$$y = 2x - 2$$

where both the independent variable x and the dependent variable y only appear to the *first power*. Let's formalize this idea in the form of a clear definition:

Definition-1: (Linear Function)

A **linear function** is any function which can be written in the form:

$$y = f(x) = mx + b \qquad (1)$$

where m and b are real numbers.

Observe that the domain of any linear function is the entire set of real numbers R, and if $m \neq 0$, the range is R as well. In the special case $m = 0$, the range is simply the set consisting of the single element b.

Example-1: Determine whether the following equations define linear functions:
(a) $2x - y = 0$
(b) $x = 2$
(c) $y = 1$

Solution:
(a) If we rewrite this equation as:

$$y = (2)x + (0)$$

we see that this is really Equation (1) where $m = 2$, and $b = 0$. So we conclude that this a linear function.

(b) Here, we see that the dependent variable y is absent so this is not a linear function.
(c) This is just the same as Equation (1) with $m = 0$ and $b = 1$, so that:

$$y = (0)x + (1)$$

We conclude that this is a linear function.
Observe that a linear function may also be written in the alternative form:

$$Ax + By = C \quad (\text{where } B \neq 0) \tag{1a}$$

where A,B and C are real numbers, and $B \neq 0$. To see this, note that since $B \neq 0$, we can solve Equation (1a) for y to give:

$$y = -\frac{A}{B}x + \frac{C}{B}$$

which has the same form as Equation (1) with $m = -A/B$ and $b = C/B$.
Now that we have defined the concept of a linear function, we turn to the operation of graphing these functions. In general, the **graph** of a function is the set of all points $P(x,y)$ in the plane whose coordinates satisfy the equation which defines the given function. We formalize this concept in Definition-2 below.

Definition-2: (Graph of a Function)

The **graph** of any function of the form: $y = f(x)$ is the set of *all* points $P(x,y)$ in the plane whose coordinates *satisfy* the equation which defines the given function.

Even though it is true that not every equation defines a function; we can still define the **graph of an equation** as the set of all points $P(x,y)$ in the xy-plane whose coordinates satisfy the given equation.
In more advanced courses, it is usually shown that *the graph of a linear function is always a straight line in the xy-plane.* Since only *one* straight line can be drawn through any two different points in the plane, *we only need two points to graph any given linear function.* Take a look at the next example.

Example-2: Construct the graph of the linear function:

$$y = -2x + 2$$

Solution:
(a) We begin by choosing *any* two convenient values for x, and then calculate the corresponding values for y. Since any two values of x will do, we choose the values $x = 0$ and $x = 1$. We then obtain:

$$\text{if } x = 0, \text{ then } y = -2(0) + 2 = 2$$
$$\text{if } x = 1, \text{ then } y = -2(1) + 2 = 0$$

(b) We now organize this information in the table below:

x	y	**Ordered Pairs**
0	2	(0,2)
1	0	(1,0)

From the table, we see that since the points (0,2) and (1,0) satisfy the equation which defines the function, then they must be on its graph. As shown below, we now plot these two points (as dots), and connect them by a straight line. This is the graph of our function:

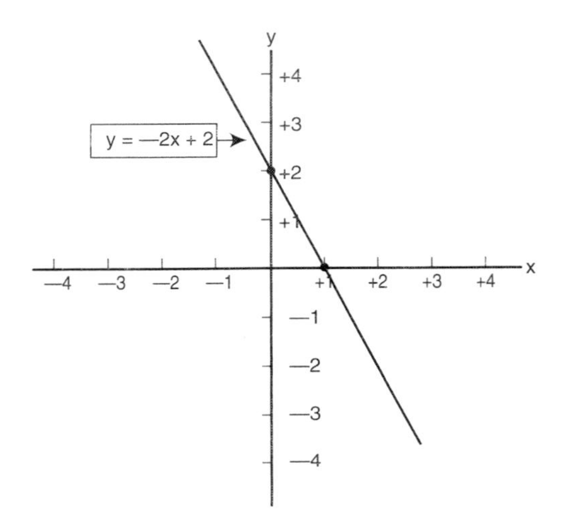

Example-2: Graph of $y = -2x + 2$

Although we have only drawn a portion of the graph, the student should understand that this line continues indefinitely in both directions.

Example-3: Graph the linear function: $y - 3x = 1$

Solution:
(a) We first solve the equation for y and get:

$$y = 3x + 1$$

(b) As before, we now choose *any* two convenient values for x, and find the corresponding values for y. Since any two values of x will do, we choose the values $x = 0$ and $x = 2$. We now substitute these values of x into the last equation:

$$\text{if } x = 0, \text{ then } y = 3(0) + 1 = 1$$
$$\text{if } x = 2, \text{ then } y = 3(2) + 1 = 7$$

(c) We now organize this information in the table below:

x	y	**Ordered Pairs**
0	1	(0,1)
2	7	(2,7)

Since the points (0,1) and (2,7) satisfy the above equation, we conclude that both points must be on its graph. We now plot these two points (as dots) and connect them by a straight line. This is what the graph looks like:

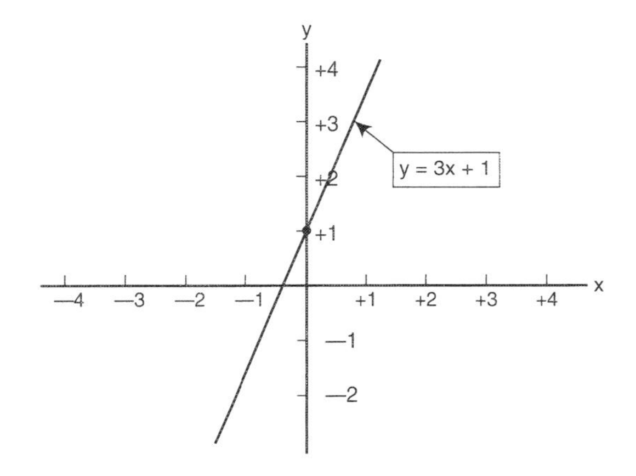

Example-3: Graph of $y = 3x + 1$

We summarize the procedure for graphing a function below:

TO GRAPH A LINEAR FUNCTION

Step #1: Solve the equation for y in terms of x.
Step #2: Choose any two values for x, say x_1 and x_2.
Step #3: Find the corresponding values for y, say y_1 and y_2.
Step #4: Plot the points (x_1, y_1) and (x_2, y_2).
Step #5: Draw a straight line through the two points.

Another way of graphing a linear function is to find its x-intercept and y-intercept. In Example-2, observe that the line $y = -2x + 2$ crosses the x-axis at the point (1,0), and crosses the y-axis at the point (0,2). We call these points the **x-intercept** and **y-intercept** of the line, respectively.

Since the y-coordinate of any point which lies on the x-axis must be zero, we can find out where the graph of any function crosses the x-axis by setting $y = 0$, and solving for x. By the same token, we know that the x-coordinate of any point on the y-axis must be zero, so we can determine where the graph of any function crosses the y-axis by setting $x = 0$, and then solving for y. This procedure is summarized below:

TO FIND THE x AND y-INTERCEPTS

To find the x-intercept: set $y = 0$ and solve the equation for x.
To find the y-intercept: set $x = 0$ and solve the equation for y.

Once we know the x and y-intercepts for a given line, we now have two points that the line must pass through, so we simply draw a line through the intercepts, and we are done.

Example-4: Draw a graph of the line $2x + 4y = 12$ by finding its x-intercept and y-intercept.

Solution:
(a) We find the x-intercept by setting $y = 0$ in the original equation, and then solving for x:

$$2x + 4y = 12 \quad \text{Original Equation}$$
$$2x + 4(0) = 12 \quad \text{Set } y = 0$$
$$x = 6 \quad \text{Solve for } x$$

We conclude that the line crosses the x-axis at the point where $x = 6$, i.e., the x-intercept is the point (6,0).
(b) We now find the y-intercept by setting $x = 0$ in the original equation, and then solving for x:

$$2x + 4y = 12 \quad \text{Original Equation}$$
$$2(0) + 4y = 12 \quad \text{Set } x = 0$$
$$y = 3 \quad \text{Solve for } y$$

So the line crosses the y-axis at the point where $y = 3$, and the y-intercept is the point (0,3). We now plot the intercepts (6,0), and (0,3) and connect them by a straight line. This is the graph of the given line.

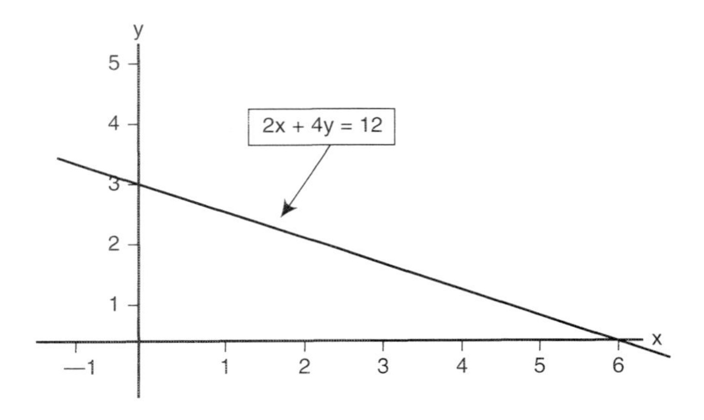

Example-4: Graph of $2x + 4y = 12$

We now turn to the concept of the **slope** or "steepness" of a straight line. This is simply a number which tells us how much a given line is inclined or slanted relative to the x-axis. We give a precise definition of this idea below:

Definition-3: (Slope of a Line)

Let $P_1(x_1,y_1)$ and $P_2(x_2,y_2)$ be *any* two points on a straight line, then the **slope** m of the line is given by:

$$m = \frac{y_2 - y_1}{x_2 - x_1} \quad \text{(provided that } x_1 \neq x_2\text{)} \qquad (2)$$

Let's try to understand the significance of this concept by drawing a simple picture. In the diagram below, we have drawn a straight line through the points P_1 and P_2 in the xy-plane.

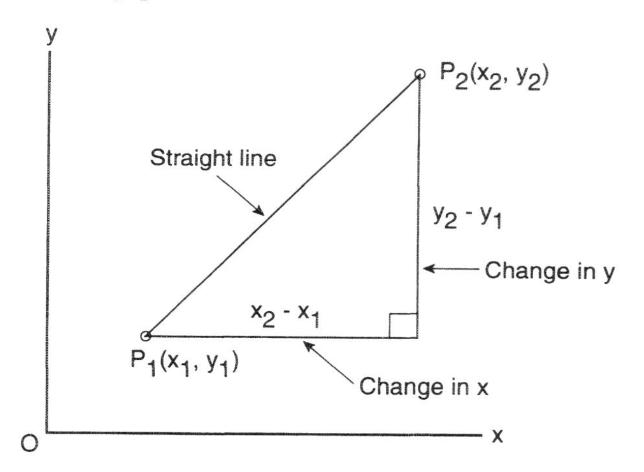

Figure 1: The Slope Concept

Observe that the expression $y_2 - y_1$ represents the change in the variable y, while $x_2 - x_1$ gives us the change in the value of x. From this diagram, and from Equation (2) above, we can write:

$$\text{slope} = \frac{y_2 - y_1}{x_2 - x_1} = \frac{\text{change in } y}{\text{change in } x} \qquad (3)$$

Thus, *the slope of a line is the ratio of the change in the dependent variable y per unit change in the independent variable x.*

Another way of writing the slope equation is to use what mathematicians call **increment notation.** The increment in any variable is just the new value of the variable minus its old value. For example, the increment in the variable y is denoted by Δy (which is read "delta y"), and is simply:

$$\Delta y = y_2 - y_1$$
$$= \text{new value of } y - \text{old value of } y$$
$$= \text{change in } y$$

and similarly, the increment in x is given by:

$$\Delta x = x_2 - x_1$$
$$= \text{new value of } x - \text{old value of } x$$
$$= \text{change in } x$$

We can now combine all of the equations for slope into a single equation for easy reference:

SLOPE FORMULA

$$\text{Slope} = \frac{y_2 - y_1}{x_2 - x_1} = \frac{\text{change in } y}{\text{change in } x} = \frac{\Delta y}{\Delta x} \tag{4}$$

Graphically, a line a has a positive slope if the line rises as we move from left to right, and its slope is negative if the line falls as we move from left to right. This situation is depicted in Figure-2 below:

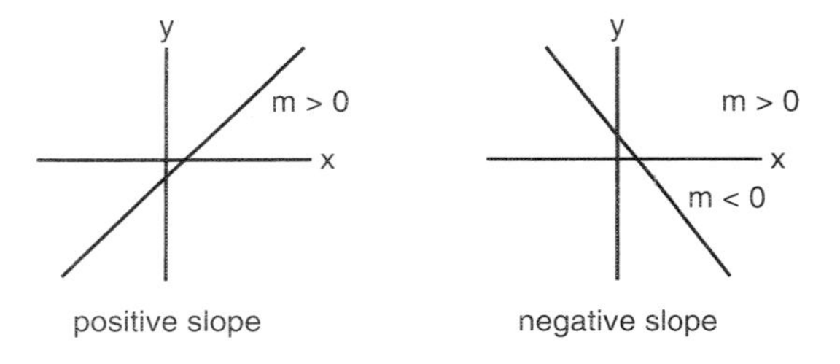

Figure 2: Lines with Positive and Negative Slopes

Now, *the slope of any line which is parallel to the x-axis is zero* since the y-coordinates of all points on such a line are all equal and therefore, $\Delta y = 0$. Consequently, the slope of a line parallel to the x-axis is:

$$m = \frac{\Delta y}{\Delta x} = \frac{(0)}{\Delta x} = 0$$

Similarly, *the slope of any line which is parallel to the y-axis is undefined.* This is the case because the x-coordinates of all points which lie on a line parallel to the y-axis must all be the same, and therefore $\Delta x = 0$. This means that the slope of such a line would be:

$$m = \frac{\Delta y}{\Delta x} = \frac{\Delta y}{(0)} \quad \text{(undefined)}$$

which is an undefined quantity since division by zero is itself undefined. These concepts are depicted in Figure-3 below.

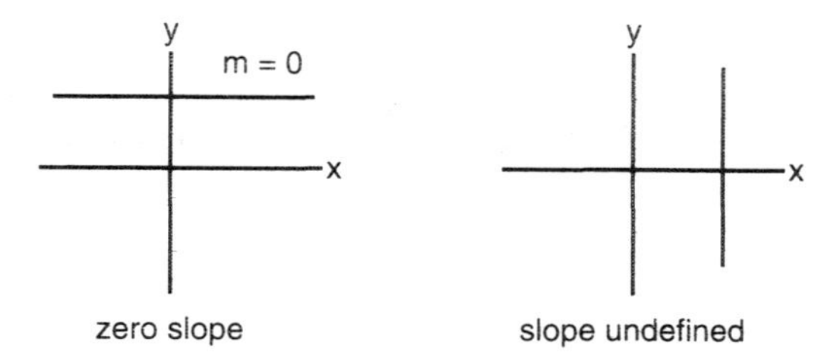

Figure 3: Lines with Zero and Undefined Slopes.

We now present a few examples which apply these ideas.

Example-5: A line passes through the points (1,−2) and (2,4). Find its slope.

Solution:
(a) We'll take (1,−2) as the first point, and (2,4) as the second point so that we have:

$$x_1 = 1, \quad y_1 = -2$$
$$x_2 = 2, \quad y_2 = 4$$

(b) By Equation (2), we get:

$$m = \frac{y_2 - y_1}{x_2 - x_1} = \frac{4-(-2)}{2-1}$$
$$= \frac{4+2}{1} = 6 \qquad \text{Solution.}$$

Example-6: On a certain day, the temperature in Miami changed from 92°F at noontime to 83°F during the evening. If T represents the temperature, then find ΔT, i.e., the change in temperature.

Solution:
(a) According to the definition of an increment, we have:

$$\Delta T = T_2 - T_1$$
$$= \text{new value of temperature} - \text{old value of temperature}$$
$$= 83°\text{F} - 92°\text{F} = -9°\text{F}$$

Note here that the negative sign indicates that the temperature was *decreasing* during the interval of time which began at noontime and ended at evening. In the event that an increment of some variable is positive, this means that this variable is increasing over the given interval.

You may have noticed that we used the symbol m for the slope of a line in Equation (2) and the same symbol for the coefficient of x in Equation (1) above. This is more than a coincidence as the next theorem will show.

Theorem-1: (Slope-Intercept Form)

If a linear function is written in the form:

SLOPE-INTERCEPT FORM

$$y = mx + b \tag{5}$$

where m and b are real numbers, then when this function is graphed to give a straight line, m is the slope of this line and $(0,b)$ is its y-intercept.

Example-7: Given the linear function $2x + 4y = 8$, find its slope and y-intercept.

Solution:
(a) First, we solve the given equation for y in order to put it into slope-intercept form:

$2x+4y=8$	Given Equation
$2x+4y-2x=8-2x$	Subtract $2x$ from both sides.
$4y=-2x+8$	Simplify
$\frac{4y}{4}=-\frac{2x}{4}+\frac{8}{4}$	Divide both sides by 4.
$y=-\frac{1}{2}x+2$	Slope - Intercept Form

(b) If we compare the last equation with equation (5), we see that the slope $m = -1/2$ and the y-intercept is (0,2).

We now conclude this lesson with a useful little theorem which enables us to determine the equation of any line if we know its slope and a point that its passes through.

Theorem-2: (Point-Slope Form)

If a straight line has slope m, and passes through the point $P_1(x_1,y_1)$, then its equation is given by:

POINT-SLOPE FORM

$$y - y_1 = m(x - x_1) \tag{6}$$

$$\frac{4y}{4} = -\frac{2x}{4} + \frac{8}{4} \quad \text{Divide both sides by 4.}$$

$$y = -\frac{1}{2}x + 2 \quad \text{Slope - Intercept Form}$$

(b) If we compare the last equation with equation (5), we see that the slope $m = -1/2$ and the y-intercept is (0,2).

We now conclude this lesson with a useful little theorem which enables us to determine the equation of any line if we know its slope and a point that its passes through.

Theorem-2: (Point-Slope Form)

If a straight line has slope m, and passes through the point $P_1(x_1,y_1)$, then its equation is given by:

POINT-SLOPE FORM

$$y - y_1 = m(x - x_1) \quad (6)$$

Example-8: A certain line has a slope of 2 and passes through the point (1,3). What is its equation?

Solution:

(a) By the previous theorem, its equation is easily determined. Here we have $m = 2$, and $(x_1,y_1) = (1,3)$. Substituting into Equation (6) we obtain:

$$y - 3 = 2(x - 1)$$

(b) Upon simplifying the last equation, we get:

$$y = 2x + 1$$

as the equation for our line.

EXERCISE SET #13

PART A

In Exercises 1–10, determine whether each statement is true or false. If a statement is false, then explain why it is false.

1. $x = 2$ is the equation of a vertical line.
2. $y = x^2 + 1$ is an example of a linear function.
3. The graphs of all linear functions are straight lines.
4. Any vertical line has a slope equal to zero.
5. The slope of the line $y = 2x + 3$ is 3.
6. The line $3x + y = 0$ passes through the origin (0,0).
7. $y = 1$ is the equation of a horizontal line.
8. Any horizontal line has a slope of zero.
9. $x = 3$ defines a linear function.
10. The line $y - 1 = 2(x + 2)$ passes through the point (2,1).

In Exercises 11–17 graph the given lines using the method outlined in Example-2, and Example-3.

11. $y = x + 2$
12. $y = -2x + 4$
13. $y = \frac{1}{2}x + 1$
14. $3x + y = 2$ [Hint: First solve for y]
15. $y - 2x = 0$
16. $y = 1$
17. $y = -4x + 2$

In Exercises 18–25, find the slope of the line which passes through the two given points. If the slope is not defined, then indicate the same.

18. (1,2) and (2,4)
19. (−1,1) and (2,3)
20. (2,1) and (10,1)
21. (1,4) and (1,−10)
22. (3,−4) and (10,2)
23. (0,1) and (4,4)
24. (1,−3) and (2,3)
25. (2,4) and (4,10)

In Exercises 26–32, find the x-intercept and y-intercept of each line, and use these intercepts to graph the line.

26. $y = x + 1$

27. $2x + y = 8$
28. $x - 2y = 4$
29. $3x + y = 6$
30. $y - 2x = 8$
31. $2x + 4y = 12$
32. $4x - 10y = 40$

In Exercises 33–39, find the slope and y-intercept of each line by first writing the equation of the line in slope-intercept form.

33. $y - 2x = 2$
34. $x + 2y = 4$
35. $4x - 2y - 4 = 0$
36. $12y - 12 = 24x$
37. $y = 3$
38. $3x + 3y = 0$
39. $2x - y - 2 = 0$

PART B

40. A line has a slope of 3 and passes through the point (0,2). Find the equation of the line.
41. A line has a slope of −2 and passes through the point (1,1). Determine the equation of the line.
42. A line passes through the points (1,2) and (3,7). Find the equation of the line. [**Hint:** First find its slope!]
43. Prove that if a line passes through the points $P_1(x_1,y_1)$ and $P_2(x_2,y_2)$, then its equation may be written in the form:

$$\frac{y - y_1}{x - x_1} = \frac{y_2 - y_1}{x_2 - x_1} \tag{7}$$

Equation (7) is called the **two-point form** of the equation for a straight line.
44. Use the two-point form of Exercise-43 to find the equation of the line which passes through the points (1,1) and (2,3).
45. Any two lines are parallel if their corresponding slopes are equal. Using this fact, find the equation of the line which passes through the point (3,3) and which is parallel to the line: $8x + 4y = 12$.

14 Introduction to Linear Systems

In many practical problems it is necessary to solve what are called **linear systems** or **systems of simultaneous equations.** In such cases, we must try to find a *single* solution, or a single pair of values for x and y which satisfies *both* equations at the same time.
For example, the linear system given by:

$$2x + 4y = 8 \qquad (1)$$
$$x + y = 3 \qquad (2)$$

has the **solution:** $x = 2$, $y = 1$ since when this pair of values is substituted for x and y in both equations, then we obtain true statements:

$$2(2) + 4(1) = 8 \qquad (1)$$

$$(2) + (1) = 3 \qquad (2)$$

In somes cases, it is possible to solve such linear systems by trial and error. A more general approach, however, is to find the **graphical solution** of linear systems. In order to see how this works, look at the next example.

Example-1: Solve the linear system given by:

$$2x + 4y = 8$$
$$x + y = 3$$

by using the graphical method.

Solution:
(a) We will graph both equations on the same set of axes. Using the intercept method, we find the following:

Equation	x-intercept	y-intercept
$2x + 4y = 8$	(4,0)	(0,2)
$x + y = 3$	(3,0)	(0,3)

(b) Using the information from the above table, we graph both lines on the same set of axes:

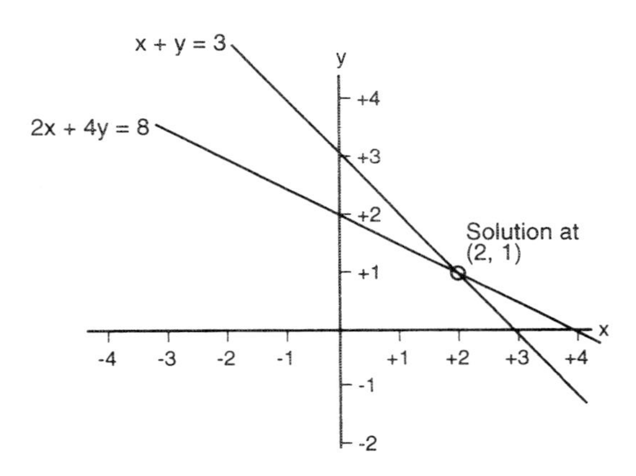

Example-1: Solution at (2,1)

(c) We observe that the point (2,1) lies on *both* lines, this means that the values $x = 2$ and $y = 1$ must satisfy both equations at the same time. Hence, our solution is $x = 2$ and $y = 1$.

Example-2: Use the graphical method to solve the linear system given by:

$$3x + 2y = 6$$
$$2x + 4y = 4$$

Solution:

(a) As before, we graph both equations on the same set of axes. Using the intercept method, we find:

Equation	x-intercept	y-intercept
$3x + 2y = 6$	(2,0)	(0,3)
$2x + 4y = 4$	(2,0)	(0,1)

(b) Based upon the above table, the graphs of these lines are as follows:

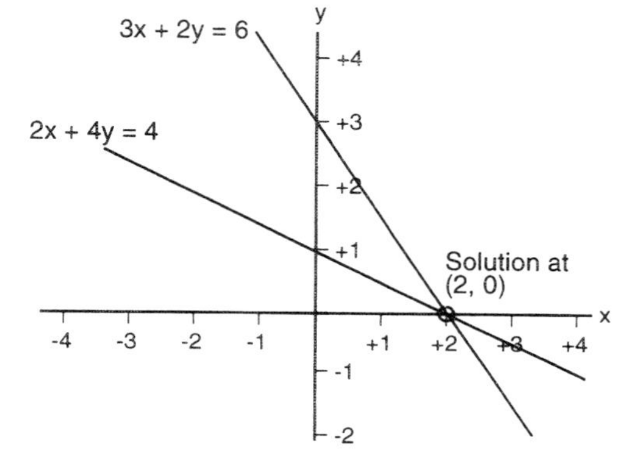

Example-2: Solution at (2,0)

(c) Notice that the point (2,0) lies on *both* lines, and consequently, its x and y coordinates must satisfy both equations simultaneously. We conclude that our solution is $x = 2$ and $y = 0$.
(d) If we wish, we can check this solution by direct substitution; this is not necessary is this case, however, since (2,0) accidently appears twice in the above table!
Observe that in both Example-1 and Example-2, each system had a *unique* solution (one and only one pair of values for x and y). Whenever a linear system has one and only one pair of values for x and y which satisfy both equations, then the system is said to be a **consistent system.** Geometrically, consistent systems are characterized by equations whose straight lines only intersect at a single point.
The next example will show that some linear systems don't have any solutions at all.

Example-3: Solve the following linear system by graphing:

$$2x + 4y = 4$$
$$4x + 8y = 16$$

(a) As in the past, we find the x and y intercepts for each line:

Equation	x-intercept	y-intercept
$2x + 4y = 4$	(2,0)	(0,1)
$4x + 8y = 16$	(4,0)	(0,2)

and draw the graphs of each line on the same set of axes:

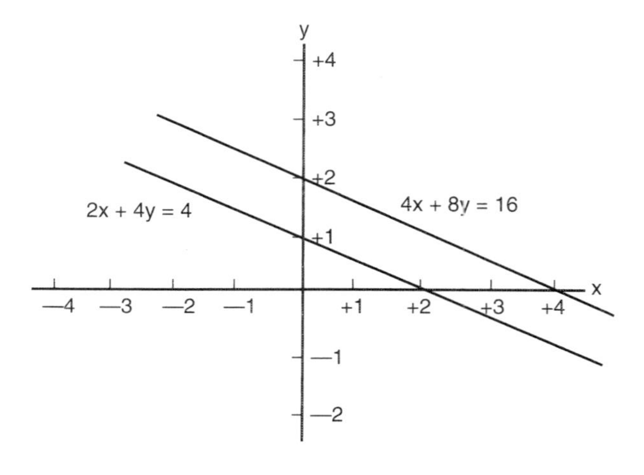

Example-3: An Inconsistent System

(b) We observe the lines are parallel so that they never intersect anywhere in the xy-plane. This means that the linear system doesn't have any solutions at all.

Whenever a linear system doesn't have any solutions, then the system is said to be an **inconsistent system.** In terms of the geometry of straight lines, this means inconsistent systems are always characterized by equations whose straight lines are parallel and never intersect.

Example-4: Use the graphical method to solve the system of equations:

$$2x + 4y = 4$$
$$4x + 8y = 8$$

Solution:

(a) If we graph boths lines by finding the x and y intercepts, then something unusual happens: the x and y intercepts for both lines are identical:

Equation	x-intercept	y-intercept
$2x + 4y = 4$	(2,0)	(0,1)
$4x + 8y = 8$	(2,0)	(0,1)

(b) Since only one line can be drawn through two given points, this means that the graphs of both lines are also the same.

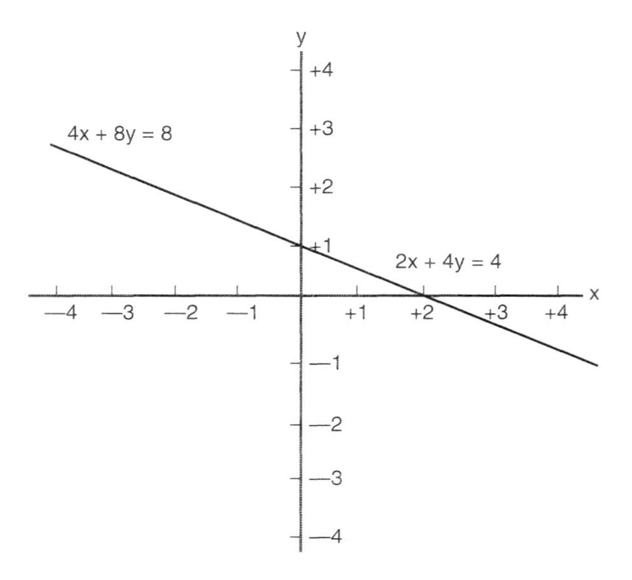

Example-4: A Dependent System

Consequently, both lines have infinitely many points in common, and *the system itself has infinitely many solutions.* Whenever a linear system has infinitely many solutions, then the system is said to be a **dependent system.** Geometrically speaking, dependent systems are characterized by equations whose graphs are identical.

The essential information in the above examples can be summarized in the possible cases shown in Figure-1 and Figure-2 below:

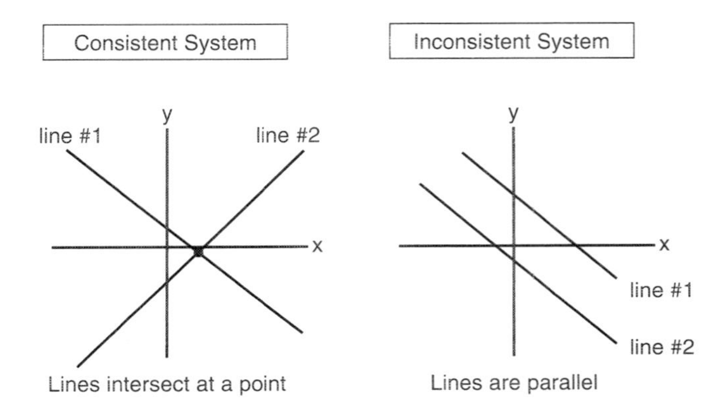

Figure 1: Consistent and Inconsistent Systems

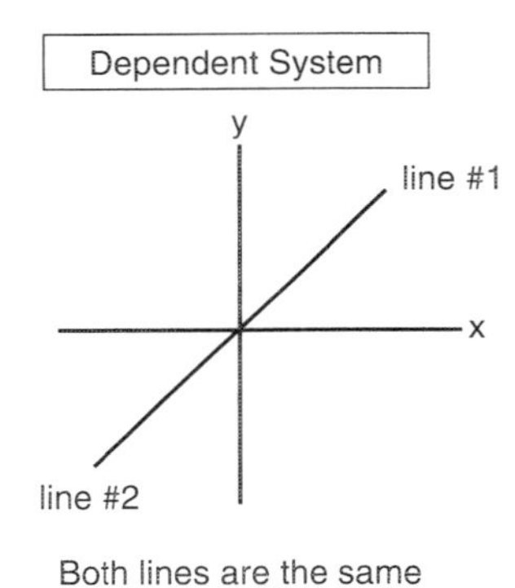

Figure 2: A Dependent System

We now complete this lesson by showing how linear systems can be used to solve everyday problems. Take a look at the last example.

Example-5: The operating costs at the Shady Rest Inn consist of a fixed daily cost of $200, and an operating cost (per occupied room) of $10 per room. The daily revenue is $50 per occupied room. Determine the following:

(a) How many rooms must be rented (on a daily basis) for the inn to breakeven?

(b) If just 8 rooms are rented, then does the inn make a profit? If so, then how much?

Solution:

(a) The total daily revenue R derived from renting x rooms is given by:

$$R = 50x$$

and the total daily cost C is given by:

$$C = 10x + 200$$

The inn will breakeven when the net profit is zero, i.e., when

$$R = C \text{ (Breakeven Point)}$$

To determine the breakeven point, we graph both the revenue and cost functions on the same set of axes and determine where they meet.

From examining the diagram below, it is apparent that the breakeven point occurs when $x = 5$ rooms are rented since the total revenue R is equal to the total cost C at that point.

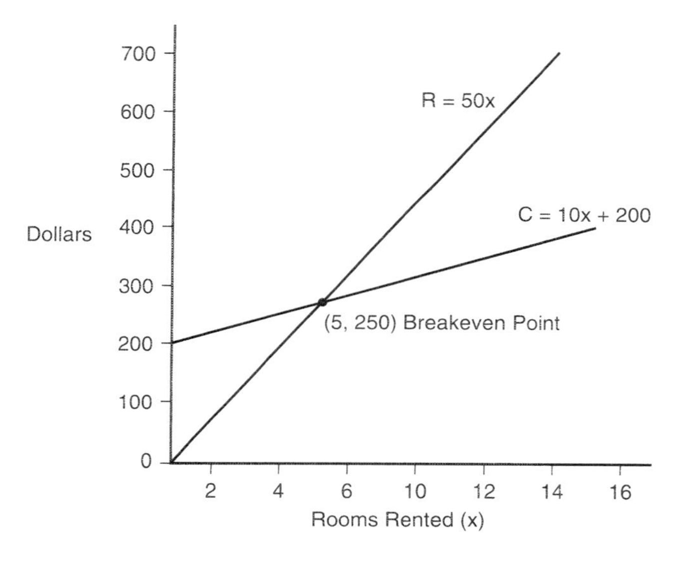

Example-5: The Shady Rest Inn

(b) If just 8 rooms are rented then the inn does make a profit since the total revenue is greater than the total cost when $x = 8$. In fact, we can find the net profit P for any value of x by:

$$\text{Net Profit} = \text{Total Revenue} - \text{Total Cost}$$
$$P = R - C$$
$$P = 50x - (10x + 200)$$
$$P = 40x - 200$$

For the particular case when $x = 8$, we have:

$$P = 40(8) - 200 = 120$$

so the inn will realize a net daily profit of \$120.

EXERCISE SET #14

PART A

In exercises 1–10, use the graphical method to solve each system of equations. State whether each system is consistent, inconsistent, or dependent.

1. $$x + y = 2$$ $$x + 2y = 3$$

2. $$x + y = 0$$ $$x + 3y = 2$$

3. $$x + y = 1$$ $$3x + 3y = -1$$

4. $$2x + y = 5$$ $$x - y = 1$$

5. $$x + 2y = 5$$ $$x + y = 3$$

6. $$-2x - 2y = 8$$ $$x + 2y = -6$$

7. $$2x - y = 0$$ $$x + y = 0$$

   Hint: The intercept method won't work for this problem, so just find two points on each line.

8. $$x + 2y = 9$$ $$2x + y = 7$$

9. $$2x + 2y = 4$$ $$2x + 2y = 6$$

10. $$x + y = -1$$ $$-2x - 2y = 2$$

In Exercises 11–15, determine whether each statement is true or false.

11. In a consistent system, the two lines must have different slopes.
12. If the two lines in a linear system have the same slope, then it must be an inconsistent system.
13. If the two lines in a linear system have the same slope, but different y-intercepts, then the system is an inconsistent system.

14. In a dependent system, both lines have the same slope and the same y-intercepts.
15. A linear system always has at least one solution.

PART B

16. A student has 85 cents in change consisting only of nickels and dimes. He has a total of 11 coins in his pocket.

    (a) If x is the number of nickels, and y is the number of dimes he has, write down a linear system (of two equations) which describes this problem.
    (b) Solve the linear system graphically to determine x and y.

17. A farmer in Texas has a total of 12 miles of fence to build a rectangular pasture for cattle. The length of the pasture is to be exactly twice its width.

    (a) If x is the length of the pasture, and y represents its width, then write down a linear system which describes this situation.
    (b) Solve the linear system graphically to determine the length and width of the farmer's pasture.

18. The daily operating costs at Joe's Family Restaurant consist of a fixed daily cost of $100, and an average meal cost of $2.50 per meal served. Joe has determined that an average price of $5.00 per meal is charged to its customers.

    (a) On the average, how many meals must be served each day for the restaurant to breakeven?
    (b) Will the restaurant make a profit if it serves 50 meals daily? If so, then how much profit?
    (c) Find the net daily profit as a function of total meals served.
    (d) What would be the net profit if 60 meals are served daily?

19. The digits in a certain two-digit number add up to 5. When its digits are reversed, the resulting number is 9 more than the original number.

    (a) Let x be the tens digit of the given number, and let y represent the units digit. This means that any two digit number can be written as $10x + y$. Write down a linear system which describes this problem.
    (b) Solve the linear system graphically to determine the unknown number.

20. The sum of two numbers is 15. The second number is one more than six times the first number.

(a) Let x be the first number, and y the second number. Write down a system of equations for this problem.
(b) Solve the linear system graphically in order to find the unknown numbers.

21. An athlete in training runs at a speed of 5 miles hour and swims at a rate of 4 miles per hour. She travels a total distance of 23 miles per day, and she runs 7 miles more than she swims each day.

(a) Let x be the miles travelled running, and y be the miles travelled swimming. Write down a linear system which describes this problem.
(b) Solve the linear system graphically to determine how far she runs and and how far she swims each day.

22. The Signet Company is a small manufacturing company that makes costume jewelry. Signet plans to make a new type of novelty ring. There is a one-time machine setup cost of \$150, and a fabrication cost of \$1.00 per ring. The selling price per ring is \$2.50.

(a) Let x be the number of rings produced and sold. Find the cost C of producing x rings.
(b) Find the revenue R resulting from the sale of x rings.
(c) Calculate the net profit P which results from the sale of x rings.
(d) How many rings must be produced and sold for the Signet Company to breakeven?

23. The sum of two numbers is 33. The larger of the two numbers is three more than twice the smaller number.

(a) If x is the smallest number, and y is the largest number, then write down a linear system of equations which describes this problem.
(b) Solve the linear system graphically to determine the unknown numbers.

24. The Shore Investment Company invested a total of \$12 million in bonds for its clients. Part of this money was invested in bonds yielding 5% a year, and the remainder of the money in bonds yielding 7% a year. After one year, the incomes derived from each type of bond were equal.

(a) Let x be the amount of money (in millions) invested in 5% bonds, and y the amount of money (in millions) invested in 7% bonds. Write down a linear system of equations which describes this problem.
(b) Solve the linear system graphically to determine how much was invested in each type of bond?

25. A shopper bought 2 shirts and one pair of trousers for a total of \$52. A week later, at the same prices, the shopper bought a total of one shirt

and 2 pairs of trousers for a total of $74. Determine the price of a shirt and the price of a pair of trousers.

26. The Sunshine Bakery sells pies at a fixed price of p per pie. The total number of pies demanded daily D is related to the price p by the equation:

$$D = -10p + 200$$

On the other hand, the daily supply of pies, S, is related to the price p per pie by the equation:

$$S = 15p - 50$$

Determine the **equilibrium price** of pies, i.e., this is the price at which the supply S and the demand D are equal.

15 More About Linear Systems

Although the graphical method for solving a linear system works well in principle, and is helpful in visualizing various problems, it still has some serious drawbacks. First, the accuracy of the graphical method is highly dependent upon the care with which we construct our graphs.
Secondly, even if we have carefully constructed a graph, it may still be impossible to come up with the exact solution to a given problem. For example, if the actual solution to a linear system is something like $x = 5/11$, and $y = 11/17$, then it may be difficult to obtain that information from your graph.

For the above reasons, **algebraic methods** have been devised for solving systems of linear equations: the **substitution method** and the **addition/subtraction method.** In this lesson, we'll present each of these methods along with several examples of their use. Let's begin with the substitution method.

SUBSTITUTION METHOD

Step #1: Select one of the two equations, and solve for one unknown variable [either x or y] in terms of the *other* variable.

Step #2: Take the expression that you derived for the unknown variable in Step #1, and substitute it into the *other* equation in the system. You should now have one equation and one unknown variable.

Step #3: Using the equation which you obtained in Step #2, solve for the single unknown variable.

Step #4: Substitute the value of the single variable you found in Step #3 into any of the original two equations and solve for the remaining unknown variable.

The above procedure may look a bit complicated at first, but the following examples should clarify this method.

Example-1: Use the substitution method to solve the linear system:

$$2x + y = 1 \quad (1)$$
$$x + 3y = 3 \quad (2)$$

Solution:
(a) We shall begin with Step #1 above. Using Equation (1), we will solve for the unknown variable y in terms of x:

$$2x + y = 1 \quad \text{Equation (1)}$$
$$2x + y - 2x = 1 - 2x \quad \text{Subtract } 2x \text{ from both sides.}$$
$$y = 1 - 2x \quad \text{Simplify.}$$

So from Equation (1), we conclude that:

$$y = 1 - 2x$$

(b) Now for Step #2, we take the expression $y = 1 - 2x$ that we derived for y in Step #1, and substitute it into the other equation:

$$x + 3y = 3 \quad \text{Equation (2)}$$
$$x + 3(1 - 2x) = 3 \quad \text{Substitute } y = 1 - 2x \text{ into (2)}$$
$$-5x + 3 = 3 \quad \text{Simplify.}$$

(c) For Step #3, we simply solve the last equation that we obtained in the previous step:

$$-5x + 3 = 3 \quad \text{From Step \#2.}$$
$$-5x = 0$$
$$x = 0 \quad \text{Solution for x.}$$

(d) For Step #4, we substitute the value $x = 0$ that we obtained into any of the original two equations and solve for the remaining variable y. If we substitute the value of $x = 0$ into Equation (1) we obtain:

$$2x + y = 1 \quad \text{Equation (1)}$$
$$2(0) + y = 1 \quad \text{Substitute } x = 0 \text{ into (1).}$$
$$y = 1 \quad \text{Solution for y.}$$

Now we are finished. We conclude that the solution of the linear system is:

$$x = 0, y = 1 \quad \textbf{Solution.}$$

Example-2: Use the substitution method to find the solution of the linear system given by:

$$x + 3y = 8 \quad (1)$$
$$2x + y = 1 \quad (2)$$

Solution:

(a) Once again, we will begin with Step #1 above. Using Equation (1), we will solve for the unknown variable x in terms of y:

$$x + 3y = 8 \quad \text{Equation (1)}$$
$$x + 3y - 3y = 8 - 3y \quad \text{Subtract } 3y \text{ from both sides.}$$
$$x = 8 - 3y \quad \text{Simplify.}$$

So from Equation (1), we have:

$$x = 8 - 3y$$

(b) Now for Step #2, we take the expression $x = 8 - 3y$ that we derived for x in Step #1, and substitute it into the remaining equation:

$$2x + y = 1 \quad \text{Equation (2)}$$
$$2(8 - 3y) + y = 1 \quad \text{Substitute } x = 8 - 3y \text{ into (2)}$$
$$-5y + 16 = 1 \quad \text{Simplify.}$$

(c) In Step #3, we simply solve the last equation that we obtained in Step #2:

$$-5y + 16 = 1 \quad \text{From Step \# 2.}$$
$$-5y = -15$$
$$\frac{-5y}{-5} = \frac{-15}{-5}$$
$$y = 3 \quad \text{Solution for y.}$$

(d) For Step #4, we substitute the value $y = 3$ that we obtained into any of the original two equations and solve for x. If we now substitute the value of $y = 3$ into Equation (1) we get:

$$x + 3y = 8 \quad \text{Equation (1)}$$
$$x + 3(3) = 8 \quad \text{Substitute } y = 3 \text{ into (1).}$$
$$x = -1 \quad \text{Solution for x.}$$

We conclude that our solution is:

$$x = -1, y = 3 \quad \textbf{Solution}$$

Example-3: (An Inconsistent System) Use the substitution method to solve the linear system:

$$x + y = 1 \qquad (1)$$
$$2x + 2y = 3 \qquad (2)$$

Solution:

(a) Starting with Equation (1), we will solve for the x in terms of in terms of y:

$$x + y = 1 \quad \text{Equation (1)}$$
$$x + y - y = 1 - y \quad \text{Subtract } y \text{ from both sides.}$$

$$x = 1 - y \quad \text{Simplify.}$$

So from Equation (1), we have:

$$x = 1 - y$$

(b) We now take the expression $x = 1 - y$ that we derived in (a) and substitute it into the remaining equation:

$$\begin{aligned} 2x + 2y &= 3 \quad \text{Equation (2)} \\ 2(1 - y) + 2y &= 3 \quad \text{Substitute } x = 1 - y \text{ into (2)} \\ 2 &= 3 \quad \text{Simplify.} \end{aligned}$$

We note, however, that since it is impossible that $2 = 3$, we are forced to conclude that *the original system of equations does not have a solution.* This type of "nonsense statement" always arises when the original system is inconsistent, i.e., the lines corresponding to each equation are parallel and have different y-intercepts.

Example-4: (A Dependent System) Use the substitution method to solve the system of equations given by:

$$x + y = 1 \qquad (1)$$
$$2x + 2y = 2 \qquad (2)$$

Solution:
(a) Starting with Equation (1), we solve for x in terms of y:

$$\begin{aligned} x + y &= 1 \quad \text{Equation (1)} \\ x &= 1 - y \quad \text{Solve for } x. \end{aligned}$$

(b) We now substitute the expression: $x = 1 - y$ into Equation (2) to get:

$$\begin{aligned} 2x + 2y &= 2 \quad \text{Equation (2)} \\ 2(1 - y) + 2y &= 2 \quad \text{Substitute } x = 1 - y \text{ into (2).} \\ 2 &= 2 \quad \text{Simplify.} \end{aligned}$$

Since the resulting statement $2 = 2$ is always true, we conclude that *the system must have infinitely many solutions,* that is, we are dealing with a **dependent system.** Whenever you arrive at an *identity* such as $0 = 0$, $1 = 1$, etc., when attempting to find the solution of a system of linear equations, you can safely conclude that the given system is dependent, and therefore it has infinitely many solutions.
Now that we have presented several examples of the substitution method, we turn to the addition/subtraction method.

ADDITION/SUBTRACTION METHOD

Step #1: If necessary, put the system of equations into **standard form,** i.e., in the form:

$$ax + by = e$$
$$cx + dy = f$$

where $a,b,c,d,e,$ and f are real numbers.

Step #2: Choose an unknown variable to eliminate. Multiply *both sides* of each equation (if necessary) by an appropriate non-zero number in order to make the coefficients (of the unknown variable that you selected) have the same *absolute value* in each equation.

Step #3: Either add or subtract the two equations you obtained in Step #2 to eliminate the unknown variable.

Step #4: Solve the equation that you obtained in Step #3 for the unknown variable.

Step #5: Substitute the value of the variable that you found in Step #4 into any of the original two equations and solve that equation for the other variable.

In the examples that follow, we will solve some of the previous problems in this lesson by using the addition/subtraction method.

Example-5: Use the addition/subtraction method to solve the linear system:

$$2x + y = 1 \qquad (1)$$
$$x + 3y = 3 \qquad (2)$$

Solution:

(a) The given system is already in standard form, so we go directly to Step #2.

(b) Let us eliminate the variable x. We notice that if we were to multiply Equation (2) by the number 2, then the coefficient of x would be the *same* in both equations:

$$2x + y = 1 \quad \text{Equation (1)}$$
$$2x + 6y = 6 \quad \text{2 Times Equation (2)}$$

(c) In Step #3, we now *subtract* the second equation from the first to give:

$$\begin{array}{r} 2x + y = 1 \\ \underline{-2x - 6y = -6} \\ -5y = -5 \end{array}$$

Notice that when we subtracted the second equation from the first equation, using the definition of subtraction, we changed the sign of each term in the second equation and then added.
(d) In Step #4, we solve the equation we obtained (in Step #3) for the unknown variable, in this case y:

$$-5y = -5 \quad \text{From Step \# 3.}$$
$$\frac{-5y}{-5} = \frac{-5}{-5} \quad \text{Divide both sides by } -5.$$
$$y = 1 \quad \text{Simplify.}$$

(e) As our final step, we substitute the value of $y = 1$ into Equation (1) and solve for x:

$$2x + y = 1 \quad \text{Equation (1).}$$
$$2x + (1) = 1 \quad \text{Substitute } y = 1.$$
$$2x = 0 \quad \text{Simplify.}$$
$$x = 0 \quad \text{Simplify.}$$

We conclude that the solution of our linear system is given by:

$$x = 0, y = 1 \ \textbf{Solution}$$

Example-6: Use the Addition/Subtraction Method to solve the linear system:

$$x + 3y = 8 \qquad (1)$$
$$2x + y = 1 \qquad (2)$$

Solution:
(a) Once again, both equations are already in standard form. We go immediately to Step #2, and in this case, we will first eliminate the variable y. We will leave the first equation alone, and multiply both sides of the second equation by -3 to obtain:

$$x + 3y = 8 \quad \text{Equation (1)}$$
$$-6x - 3y = -3 \quad \text{Multiply Equation (2) by } -3.$$

(b) In Step #3, we will now add both equations obtained in the previous step to obtain:

$$\begin{array}{r} x + 3y = 8 \\ -6x - 3y = -3 \\ \hline -5x \quad = 5 \end{array}$$

(c) In Step #4, we solve the equation obtained in the previous step for x:

$$-5x = 5 \quad \text{From Step \# 3.}$$
$$\frac{-5x}{-5} = \frac{5}{-5} \quad \text{Divide both sides by } -5.$$
$$x = -1 \quad \text{Simplify.}$$

(d) In our last step, we will substitute the value of $x = -1$ into Equation (2), and then solve for y:

$$2x + y = 1 \quad \text{Equation (2).}$$
$$2(-1) + y = 1 \quad \text{Substitute } x = -1 \text{ into (2).}$$
$$-2 + y = 1 \quad \text{Simplify.}$$
$$y = 3 \quad \text{Solve for } y.$$

We now have the solution to our system of equations:

$$x = -1, \quad y = 3 \ \textbf{Solution.}$$

Example-7: Use the Addition/Subtraction Method to solve the linear system given by:

$$2x + 3y = -5 \qquad (1)$$
$$5x + 2y = 4 \qquad (2)$$

Solution:
(a) In this case, we will eliminate the variable x. In order to do this, we will multiply the first equation by 5, and the second equation by -2 to give:

$$10x + 15y = -25 \quad \text{Multiply Equation (1) by 5.}$$
$$-10x - 4y = -8 \quad \text{Multiply Equation (2) by } -2.$$

(b) We now add both equations together in order to eliminate x, and then we'll solve for y:

$$\begin{array}{r} 10x + 15y = -25 \\ \underline{-10x - 4y = -8} \\ 11y = -33 \end{array}$$
$$\frac{11y}{11} = \frac{-33}{11}$$
$$y = -3$$

(c) Finally, we substitute $y = -3$ into Equation (2) and solve for x:

$$5x + 2y = 4 \quad \text{Equation (2).}$$
$$5x + 2(-3) = 4 \quad \text{Substitute } y = -3.$$
$$5x - 6 = 4 \quad \text{Simplify.}$$

$$5x = 10 \quad \text{Simplify.}$$
$$x = 2 \quad \text{Solve for } x.$$

We conclude that the solution to the system is:

$$x = 2, y = -3 \quad \textbf{Solution}$$

Observe that in the previous example, we could have multiplied the first equation by 5, and the second equation by 2, and then *subtracted;* we still would have eliminated x. This is a matter of taste, as long as we adhere to the basic properties of equations then we can do just about whatever we please!

Example-8: Use the Addition/Subtraction Method to solve the linear system of equations given by:

$$3y - \frac{1}{2}x = 4 \qquad (1)$$

$$3x - \frac{1}{3}y = 0 \qquad (2)$$

Solution:
(a) We first put these equations into standard form:

$$-\frac{1}{2}x + 3y = 4 \qquad (1)$$

$$3x - \frac{1}{3}y = 0 \qquad (2)$$

and to make things easier to visualize, we shall eliminate all fractions:

$$-x + 6y = 8 \quad \text{Multiply (1) by 2.} \qquad (1a)$$
$$9x - y = 0 \quad \text{Multiply (2) by 3.} \qquad (2a)$$

(c) We observe that if we multiply Equation (2a) by 6, and then add the two equations, y will be eliminated. Let's follow this game plan:

$$\begin{array}{r} -x + 6y = 8 \\ \underline{54x - 6y = 0} \\ 53x = 8 \end{array}$$

$$\frac{53x}{53} = \frac{8}{53}$$

$$x = \frac{8}{53}$$

(d) Finally, we substitute $x = 8/53$ into Equation (2a) and then solve for y:

$$9x - y = 0 \text{ Equation(2a)}$$

$$9(\frac{8}{53}) - y = 0 \text{ Substitute } x = \frac{8}{53} \text{ into (2a).}$$

$$\frac{72}{53} - y = 0 \text{ Simplify.}$$

$$y = \frac{72}{53} \text{ Solve by y.}$$

The solution to our system is therefore:

$$x = \frac{8}{53}, \; y = \frac{72}{53} \qquad \textbf{Solution}$$

Observe that the solution to the last example would have been very difficult to obtain using the graphical method of the previous lesson. This was, after all, one of the reasons for introducing our two new methods.

EXERCISE SET #15

PART A

In Exercises 1–10, use the Substitution Method to solve each linear system. If the system is inconsistent or dependent then indicate the same.

1. $$\begin{aligned} x + y &= 0 \\ 2x + 3y &= 1 \end{aligned}$$

2. $$\begin{aligned} 2x + y &= 1 \\ 3x + 2y &= 2 \end{aligned}$$

3. $$\begin{aligned} -2x + y &= -1 \\ 3x - 2y &= 0 \end{aligned}$$

4. $$\begin{aligned} 2x + 2y &= 1 \\ 4x + 4y &= 3 \end{aligned}$$

5. $$\begin{aligned} 5x - 4y &= 9 \\ 2x - 2y &= 2 \end{aligned}$$

6. $$\begin{aligned} x + y &= 0 \\ 2x + 2y &= 0 \end{aligned}$$

7. $$\begin{aligned} 2x + 3y + 10 &= 0 \\ 3x + y &= -8 \end{aligned}$$

8. $$\begin{aligned} 2x + 3y &= 2 \\ 4x - 6y &= 0 \end{aligned}$$

9. $$\begin{aligned} 4x + 3y &= -2 \\ 12x - 6y &= 7 \end{aligned}$$

10. $$\begin{aligned} \frac{1}{2}x + \frac{1}{3}y &= 1 \\ 6x + 4y &= 12 \end{aligned}$$

In Exercises 11–20, use the Addition/Subtraction Method to solve each linear system. If the system is inconsistent or dependent then indicate the same.

11. $$\begin{aligned} 2x + 2y &= 0 \\ 2x + 3y &= 1 \end{aligned}$$

12. $$\begin{aligned} 2x + 2y &= 2 \\ 3x - 2y &= 2 \end{aligned}$$

13. $$\begin{aligned} -2x + y &= -1 \\ -2x - 2y &= 0 \end{aligned}$$

14. $$\begin{aligned} 2x + 2y &= 1 \\ 4x + 4y &= 0 \end{aligned}$$

15. $$\begin{aligned} 5x - 4y &= 9 \\ 2x + 2y &= 1 \end{aligned}$$

16. $$\begin{aligned} -2x + y &= 0 \\ 2x + 2y &= 4 \end{aligned}$$

17. $$\begin{aligned} 2x + 3y &= -3 \\ 3x + y &= -1 \end{aligned}$$

18. $$\begin{aligned} 2x + 3y &= -1 \\ 4x - 6y &= -10 \end{aligned}$$

19. $$\begin{aligned} 4x + 3y &= -2 \\ x - 2y &= 7 \end{aligned}$$

20. $$\begin{aligned} \frac{1}{3}x - \frac{1}{2}y &= 0 \\ 2x - y &= 1 \end{aligned}$$

PART B

21. A student has $1.25 in change consisting only of nickels and dimes. He has a total of 15 coins in his pocket. How many dimes and nickels does the student have?
22. A gardener has a total of 120 feet of fence to enclose her garden. The length of the garden is to be exactly twice its width. What are the dimensions of the garden?
23. The digits in a certain two-digit number add up to 7. When its digits are reversed, the resulting number is 1 less than one-half the original number. What is the original number?
24. The sum of two numbers is 25. The second number is one more than the first number. What are the numbers?
25. A music lover purchased 2 CD's and one cassette tape at a local music store for a total of $28. A day later, she returned to buy 3 cassette tapes and 1 CD for a total price of $34. If all cassette tapes are the same price, and all CD's are priced the same, then determine the price of a cassette tape, and a CD.
26. The Crown Hotel has a total of 124 rooms. There are three times as many double-rooms as single-rooms. How many rooms are there of each kind?

16 EMPIRICAL PROBABILITY

From your experience in life, you know that there is an element of unpredictability in many events that take place around us. The weather, fluctuations of the stock market, whether we win or lose at the casino, and even the behavior of subatomic particles are examples of this uncertainty. In order to handle such uncertain events, mathematicians have developed the **theory of probability.** This is an extremely important branch of mathematics that has a broad area of application in many different fields.

In order to understand just what probability means, consider the **experiment** which involves tossing a penny 1,000 times. We recorded the results of this experiment at its various stages, and results are shown below in Table-1:

TABLE-1: FLIPPING A PENNY 1,000 TIMES

Number of Tosses (N)	Number of Heads (h)	Number of Tails (t)	Relative Frequency of Heads (h/N)
100	43	57	$\frac{43}{100} = 0.430$
500	246	254	$\frac{246}{500} = 0.492$
1000	498	502	$\frac{498}{1000} = 0.498$

For example, according to Table-1, after just 100 coin tosses, we obtained 43 heads, and 47 tails; the *ratio* of heads to the total number of coin tosses is called the **relative frequency** of "heads." Notice that as the number of coin tosses increases, the relative frequency of heads gets closer and closer to 1/2.

This is just what we would expect if the coin is perfectly balanced. Since there are only two possible **outcomes** for this experiment: either heads or tails, we would expect to observe heads half of the time, and tails the other half.

In general, an **experiment** is any operation of finite duration which yields a set of results that are called the **outcomes** of the experiment. Similarly, an **event** is simply either a single outcome or any combination of outcomes. We can generalize these ideas to handle any experiment, and the result appears in the definition below:

Definition-1 (Empirical Probability)

If an experiment is repeated a total of N times, and a given event E is observed to occur a total of n times, then we say the **empirical probability** of the event E, denoted by $P(E)$, is the relative frequency (n/N) that the event occurs, i.e.,

$$P(E) = \frac{n}{N} \tag{1}$$

or in other words,

$$P(E) = \frac{\text{Number of Times Event Occurs}}{\text{Total Number of Times We Perform the Experiment}} \tag{2}$$

Strictly speaking, we say that the empirical probability is given by Equation-(1) if the relative frequency approaches a fixed number $P(E)$ as the number of times N that we perform the experiment gets larger and larger. For example, in the experiment which consisted of flipping a penny 1,000 times, observe that the ratio of the number of heads to the total number of tosses approached 1/2.

Example-1: In a survey of 9000 students at Johnson & Wales University, it was determined that 4,400 of these students were male, while 4,600 students were female.
(a) What is the probability P(male) that a student is male?
(b) What is the probability P(female) that a student is female:?

Solution:
(a) In this case, we can write an equation which is similar to Equation (2):

$$\begin{aligned} P(\text{Male}) &= \frac{\text{Number of Times "Male" Occurs in the Survey}}{\text{Total Number of Students Surveyed}} \\ &= \frac{4400}{9000} \approx 0.49 \end{aligned}$$

where the symbol "≈" means "is approximately equal to."
(b) Similarly,

$$\begin{aligned} P(\text{Female}) &= \frac{\text{Number of Times "Female" Occurs in the Survey}}{\text{Total Number of Students Surveyed}} \\ &= \frac{4600}{9000} \approx 0.51 \end{aligned}$$

Example-2: A single die is tossed 600 times, and the number which appears on its top is recorded. The results are as follows:

RESULTS OF TOSSING A SINGLE DIE (600 TIMES)	
Number on Top	**Number of Times Observed**
1	102
2	94
3	101
4	102
5	100
6	101

If the die is then tossed one more time,

(a) What is the probability P(getting a 6) of getting a six?
(b) What is the probability P(getting a 5) of getting a five?
(c) What is the probability P(getting a 2) of getting a two?

Solution:
Once again, we use Equation (2) to obtain:

(a)

$$P(\text{getting a } 6) = \frac{\text{number of times "6" occurs}}{\text{number of times experiment is performed}}$$
$$= \frac{101}{600} \approx 0.168$$

(b)

$$P(\text{getting a } 5) = \frac{\text{number of times "5" occurs}}{\text{number of times experiment is performed}}$$
$$= \frac{100}{600} \approx 0.167$$

(c)

$$P(\text{getting } a \ 2) = \frac{\text{number of times "2" occurs}}{\text{number of times experiment is performed}}$$
$$= \frac{94}{600} \approx 0.157$$

Example-3: During an eight hour period, a total of 102 customers were served at a restaurant, and of that number, 41 ordered daily specials. If the next customer were to enter the restaurant, determine the empirical probability that:

(a) the customer orders a daily special.
(b) the customer does not order a daily special.

Solution:
(a) Using the definition of empirical probability, we have:

$$P(\text{order special}) = \frac{\text{no. of customers who ordered daily special}}{\text{total number of customers}} = \frac{41}{102} \approx 0.402$$

(b) Similarly, we have:

$$P(\text{not order special}) = \frac{\text{no. of customers who didn't order daily special}}{\text{total number of customers}} = \frac{102-41}{102} = \frac{61}{102} \approx 0.598$$

EXERCISE SET #16

PART A

1. According to government statistics, in 1995 there were a total of 11,723 traffic fatalities involving single vehicles. It was also reported that about 5,400 of the drivers were driving while intoxicated.

   (a) Assuming that a single-vehicle traffic fatality has taken place, what is the probability *P*(drunk) that the driver was intoxicated.
   (b) Assuming that a single-vehicle traffic fatality has taken place, what is the probability *P*(not drunk) that the driver was not drunk?

2. In a public opinion poll, 400 people were asked to rate a Governor's overall performance as excellent, good, fair, and poor. The results were as follows:

Possible Rating	Number of People
Excellent	80
Good	200
Fair	100
Poor	20

   Assuming that this survey was representative of of the state's population, then find the probability that the next person surveyed:

   (a) rates the Governor's performance as poor.
   (b) rates the Governor's performance as excellent.
   (c) rates the Governor's performance as good.

3. Willie Mays, a member of the Baseball Hall of Fame, and who played 22 years for the San Francisco Giants, got a total of 3,283 hits and 660 homeruns in a total of 10,881 times at bat. If Willie were to bat again, then determine:

   (a) the probability that he gets a hit.
   (b) the probability that he hits a homerun.

4. An experimental drug was developed and tested on a total of 1,200 patients suffering from cancer. A total of 240 patients responded to the drug and improved dramatically. Assuming that the drug is administered to the next cancer patient, determine:

(a) the probability that the patient responds to the drug.
(b) the probability that the patient does not respond to the drug.

5. Gracie's Bistro, a small Italian restaurant, conducted a survey of 200 of its customers which rated the quality of the food it served as excellent, good, fair, poor, and no opinion. The results of the survey were as follows:

SURVEY RESULTS FOR GRACIE'S BISTRO	
Food Rating	**Number of Customers**
Excellent	104
Good	60
Fair	20
Poor	10
No Opinion	6

If the next customer were to enter the restaurant, then determine:

(a) the probability that the customer would rate the food to be excellent.
(b) the probability that the customer would rate the food to be good.
(c) the probability that the customer wouldn't have any opinion as to food quality.

6. An Olympic Sharpshooter fired at a target a total of 500 times, and hit the target a total of 480 times. If he were to fire again, then what is the probability that he would:

(a) Hit the target.
(b) Miss the target.

7. A student who enjoys playing Keno played the same combination of numbers for a total of 120 games. In 18 of these games, the student won money, and in the remainder of the games, the student lost money. If the student were to play Keno (using the same numbers) one more time, then determine the probability that the student would:

(a) win
(b) lose

8. According to records at the Baseball Hall of Fame, Mickey Mantle, who played for the New York Yankees for 18 years and who was inducted into the Hall of Fame in 1974, had the following statistics:

PLAYER STATISTICS FOR MICKEY MANTLE	
At Bats	8,102
Hits	2,415
Doubles	344
Triples	72
Home Runs	536
Walks	1,734
Strike Outs	1,710

If Mickey were to bat one more time, then determine the following:

(a) the probability that he would hit a home run.
(b) the probability that he would get a hit.
(c) the probability that he would hit a triple.
(d) the probability that he would walk.

9. The Puritan Hotel surveyed 500 of its customers to determine how they generally felt about the hotel services:

PURITAN HOTEL SURVEY	
Service Rating	**Number of Customers**
Excellent	50
Good	255
Fair	120
Poor	75

If this sample is representative of the entire population of their customers, then find the probability that:

(a) a customer rates hotel services as fair.
(b) a customer rates hotel services as good.
(c) a customer rates hotel services as poor.

10. A tetrahedron (this is a regular solid with four sides) has its faces numbered one through four as shown below:

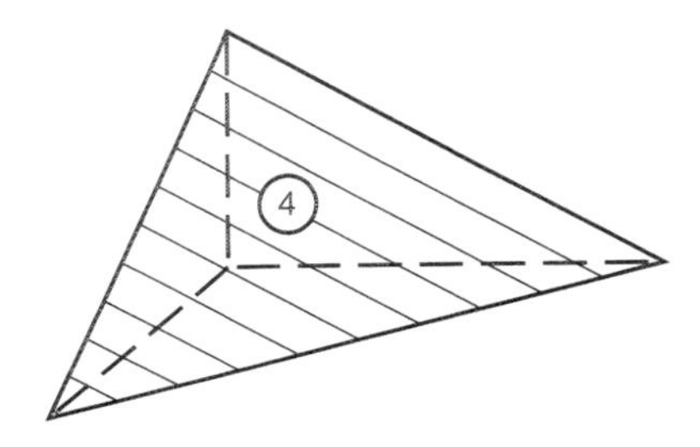

Exercise-10. A Tetrahedral Die

The tetrahedron was tossed 400 times, and the number which appeared on its bottom was recorded. The results of this experiment were as follows:

Tossing a Tetrahedral Die	
Number Recorded	**Frequency of Occurrence**
1	118
2	96
3	84
4	102

If this "tetrahedral die" were to be tossed one more time, then determine the probability of:

(a) obtaining a 4.
(b) obtaining a 3.
(c) obtaining a 1.

17 THEORETICAL PROBABILITY

In calculating the empirical probability of an event, we *physically* perform some experiment such as tossing a die 400 times, taking a survey, or flipping a coin 1000 times. We record the frequency with which the event of interest occurs, and we take the empirical (experimental) probability of that event to be its *relative frequency* of occurrence.

In some cases, however, it is possible to calculate the probability of a given event without performing any *physical* experiment at all. This happens when all of the outcomes of a given *theoretical* experiment are **equally likely,** i.e., when each outcome has the same probability of occuring as any other outcome. This type of thinking leads to the concept of **theoretical probability.**

Before defining theoretical probability in a precise way, we need some definitions.

Definition-1: (Sample Space)

The set S of all possible outcomes of any given experiment is called the **sample space** for that experiment. If S has a finite number of elements, then it is said to be a **finite sample space.**

Definition-2: (Event & Outcome)

An **outcome** of any experiment is simply an element of the sample space S for that experiment. An **event** is any subset of S. Thus, an event is any combination of one or more outcomes.

Example-1: A single die is tossed once, and we observe the number which appears on top.

(a) What is the sample space for this experiment?
(b) Describe the event $A = \{2,4,6\}$
(c) Describe the event $B = \{1,2,3\}$

Solution:
(a) The sample space S is the set of all possible outcomes for a given experiment. In this case, we can obtain a number from one to six, so:

$$S = \{1,2,3,4,5,6\}$$

is our sample space. Since it contains finitely many elements, i.e., since $\#(S) = 6$, it is also a finite sample space.
(b) Note that $A = \{2,4,6\}$ is an event since $A \subseteq S$. We could describe this event by saying that it is the "event consisting of observing an even number."
(c) Similarly, $B = \{1,2,3\}$ is also an event since $B \subseteq S$ as well. This event might be described as the "event which consists of observing a number less than four."

Example-2: A bag contains raffle tickets that are numbered from 1 to 500. A single raffle ticket is drawn at random from the bag.

(a) What is the sample space S for this experiment.
(b) Describe the event $E = \{5\}$.

Solution:
(a) The sample space S is the set of all possible outcomes for the given experiment. In this case, we can obtain a number from one to 500, so we have:

$$S = \{1,2,3,...,500\}$$

as our sample space. Since it contains finitely many elements, i.e., since $\#(S) = 500$, it is also a finite sample space.
(b) We see that E is an event since:

$$E = \{5\} \subseteq S$$

and this event consists of selecting the raffle ticket numbered five.

Now that we have captured the concepts of sample space, and event, we can proceed with our long awaited definition for theoretical probability:

Definition-3 (Theoretical Probability)

Let an experiment have a *finite* sample space S which consists of N (equally-likely) outcomes, and let the event E consist of n elements, then the probability of the event E, denoted by $P(E)$ is:

$$P(E) = \frac{n}{N} \quad (1)$$

or in other words,

$$P(E) = \frac{\text{Number of Outcomes Favorable to } E}{\text{Total Number of Possible Outcomes}} \quad (2)$$

Example-3: A single die is tossed and the number which appears on top is observed.

(a) What is the probability of observing a 3?
(b) What is the probability of observing a number which is greater than 4?
(c) What is the probability of observing an odd number?
(d) What is the probability of observing a number less than one?
(e) What is the probability of observing a number from one to six?

Solution:
From Example-1 above, we know that the sample space for this experiment is:

$$S = \{1,2,3,4,5,6\}$$

so $\#(S) = 6$. We need this fact to be able to calculate the various probabilities.
(a) Let $A = \{3\}$ be the event which consists of observing a 3. Then since $\#(A) = 1$, we have:

$$P(\text{observing a } 3) = P(A) = \frac{\#(A)}{\#(S)} = \frac{1}{6}$$

(b) Let $B = \{5,6\}$; this is the event of observing a number greater than four. Since $\#(B) = 2$, we have:

$$P(\text{observing a number} > 4) = P(B) = \frac{\#(B)}{\#(S)} = \frac{2}{6} = \frac{1}{3}$$

(c) Let $C = \{1,3,5\}$; this is the event which consists of observing an odd number. Since $\#(C) = 3$, we have:

$$P(\text{observing an odd number}) = \frac{\#(C)}{\#(S)} = \frac{3}{6} = \frac{1}{2}$$

(d) Let D be the event which consists of obtaining a number less than one. Since this is not possible, we must have $D = 0$ i.e., D is an empty set, and since an empty contains no elements, we must have $\#(D) = 0$. Consequently,

$$P(\text{observing number} < 1) = \frac{\#(D)}{\#(S)} = \frac{0}{6} = 0$$

For this reason, we call any event E where $E = \emptyset$ an **impossible event,** i.e., it can't happen.

(e) Finally, let E be the event which consists of observing a number from one to six. Then clearly, $E = \{1,2,3,4,5,6\}$ so that $\#(E) = 6$. We conclude that:

$$P(\text{observing a number from 1 to 6}) = \frac{\#(E)}{\#(S)} = \frac{6}{6} = 1$$

In general, any event E where $E = S$ is called a **certain event,** i.e., it has to happen, since for such an event, $P(E) = 1$.

Example-4: A single card is drawn at random from a standard deck of 52 cards.
(a) What is the probability of drawing a three of hearts?
(b) What is the probability of drawing an ace?

Solution:
We note that there are 52 possible outcomes for this experiment since there 52 different cards in a standard deck; and so:

$$\#(S) = 52$$

Since the card is drawn at random, then each card has the same chance of being drawn, and consequently, the outcomes are equally likely.
(a) Let A be the event which consists of drawing a three of hearts. Since there is only one such card, we have $\#(A) = 1$. We then have:

$$P(\text{Three of Hearts}) = \frac{\#(A)}{\#(S)} = \frac{1}{52}$$

(b) Let B be the event which consists of drawing an ace, then:

$$B = \{\text{ace of spades, ace of diamonds, ace of hearts, ace of clubs}\}$$

and consequently, we have $\#(B) = 4$, and:

$$P(\text{drawing an ace}) = \frac{\#(B)}{\#(S)} = \frac{4}{52} = \frac{1}{13}$$

After working through these examples, you have probably deduced that theoretical probability of any event possesses certain properties which remain the same from problem to problem. These properties have been summarized in our **probability axioms** below:

Definition-4: (Axioms for Probability)

Let S be a finite sample space that is associated with an experiment, and suppose that:

$$S = \{s_1, s_2, s_3, \ldots, s_N\}$$

Definition-4 (Continued)

Then we assume that the following axioms hold true:

(1) If E is any event, i.e., $E \subseteq S$, then:

$$0 \leq P(E) \leq 1 \qquad \text{Axiom-1}$$

In other words, the probability of any event is always a number between 0 and 1.

(2) The probability of a **certain event,** i.e., an event which is certain to occur, is one:

$$P(S) = 1 \qquad \text{Axiom-2}$$

(3) The probability of an **impossible event** is zero, i.e.,

$$P(\emptyset) = 0 \qquad \text{Axiom-3}$$

(4) The sum of the probabilities of *all* possible outcomes in the sample space S is always one:

$$P(s_1) + P(s_2) + P(s_3) + \ldots + P(s_N) = 1 \qquad \text{Axiom-4}$$

Although we have stated these axioms for a finite sample space, properties (1)–(3) can be shown to hold for any sample space. We will now look at complementary events.

Definition-5: (Complementary Events)

Let S be a finite sample space, and let E be any event, i.e., $E \subseteq S$. Then the event E' is defined to be the collection of all those elements in S that are not in the set E; in other words,

$$E' = S - E \qquad (3)$$

We call E' the **complement** of E (and vice versa), and we say that E' and E are **complementary events.**

From this definition, it is clear that *if the event E occurs, then the event E' cannot occur,* and conversely. Let's look at an example which will help to clarify this idea.

Example-6: A single die is tossed once, and the number on its top is observed. Find the complements of the following events:
(a) The event A which consists of observing an even number.
(b) The event B which consists of observing a number greater than 4.

Solution:
(a) Since A is the event which consists of observing an even number, then we have:

$$A = \{2,4,6\}$$

Since the sample space for this experiment is given by:

$$S = \{1,2,3,4,5,6\}$$

then by (3), the complement of this event is simply:

$$\begin{aligned} A' &= S - A \\ &= \{1,2,3,4,5,6\} - \{2,4,6\} \\ &= \{1,3,5\} \end{aligned}$$

We see that A' is simply the event that A itself does not happen, i.e., an odd number is observed.
(b) Similarly, since $B = \{5,6\}$, then using (3), we have:

$$\begin{aligned} B' &= S - B \\ &= \{1,2,3,4,5,6\} - \{5,6\} \\ &= \{1,2,3,4\} \end{aligned}$$

Theorem-1: (Complementary Events)

If E and E' are complementary events in a finite sample space S, then:

$$P(E) + P(E') = 1$$

or alternatively,

$$P(E) = 1 - P(E') \qquad (4)$$

In other words, *the sum of the probability that an event occurs, and the probability that the event does not occur is always one.*

In order see how useful this theorem can be, take a look at the following examples.

Example-7: A card is randomly selected from a standard deck of 52 playing cards. If one card is selected, then what is the probability that the card selected is *not* an ace?

Solution:
(a) Let E represent the event of selecting a ace. Then, we have:

$$P(E) = \frac{\#(E)}{\#(S)} = \frac{4}{52} = \frac{1}{13}$$

(b) From Definition-5, we see that E' must be the event which consists of not selecting an ace. Consequently,

$$P(E') = 1 - P(E) \text{ or, in other words,}$$
$$P(\text{not an ace}) = 1 - P(\text{ace}) = 1 - \frac{1}{13} = \frac{12}{13}$$

Example-8: It is known that 3 out of every 100 lights bulbs that are shipped to a certain retail store are defective. If a customer were to select a light bulb at random, then what is the probability that it is *not* defective?

Solution:
(a) If we let E represent the event that a light bulb is defective, then:

$$P(E) = P(\text{defective}) = \frac{3}{100}$$

(b) By the previous definition, E' must represent the event which consists of *not* finding a defective light bulb, so by Equation (4), we have:

$$\begin{aligned} P(\text{not defective}) &= 1 - P(\text{defective}) \\ &= 1 - \frac{3}{100} \\ &= \frac{97}{100} \end{aligned}$$

Exercise Set #17

PART A

In Exercises 1–10, a single card is selected at random from a standard deck of 52 playing cards. Find the probability that the card selected is:

1. a queen.
2. a jack.
3. a ten.
4. a red card.
5. a black card.
6. a face card (i.e., a jack, queen, or king).
7. a two.
8. the ace of clubs.
9. a card greater than 2 but less than 5.
10. a red face card.
11. In a certain restaurant, the management has determined that customers are extremely satisfied with the quality of their food 95 percent of the time. If a customer is selected at random, what is the probability that the customer is not extremely satisfied with the food quality?
12. A bin contains 3 red balls, 2 yellow balls, and 5 orange balls. A single ball is randomly selected from the bin. Write down the sample space S for this experiment, and find the probability of:

    (a) selecting a red ball.
    (b) selecting a yellow ball.
    (c) selecting an orange ball.
    (d) not selecting a red ball.
    (e) not selecting an orange ball.

13. A certain mathematics class contains 10 Hotel management majors, 5 Culinary majors, and 15 Business majors. If a student is randomly selected from this class, find the probability that that the student is:

    (a) a hotel major.
    (b) a culinary major.
    (c) a business major.
    (d) not a hotel major.
    (e) not a business major.

14. A shipment of 100 television sets is known to contain two defective sets. If a television set is randomly selected from the shipment, then find the probability that it is:

    (a) a defective set.
    (b) not a defective set.

PART B

15. An experiment consists of flipping a single coin two times, and observing whether you get a heads or a tails each time it is tossed. The sample space S for this experiment may be written as:

$$S = \{(H,H), (H,T), (T,H), (T,T)\}$$

where the ordered pairs are used to record what happened on each toss. For example, the ordered pair (H,T) means that we observe a heads on the first toss, and a tails on the second toss. Find the following probabilities:

(a) the probability of 2 heads.
(b) the probability of 0 heads.
(c) the probability of 1 head.
(d) the probability of *at least* 1 head.

16. A square target which is 2 feet long and 2 feet high is painted on the side of a barn. The barn's side measures 20 feet by 10 feet. A blindfolded student is asked to throw a ball at the barn. If we assume that the student does not miss the side of the barn entirely, then determine the following probabilities:

(a) the probability of hitting the target.
(b) the probability of missing the target.

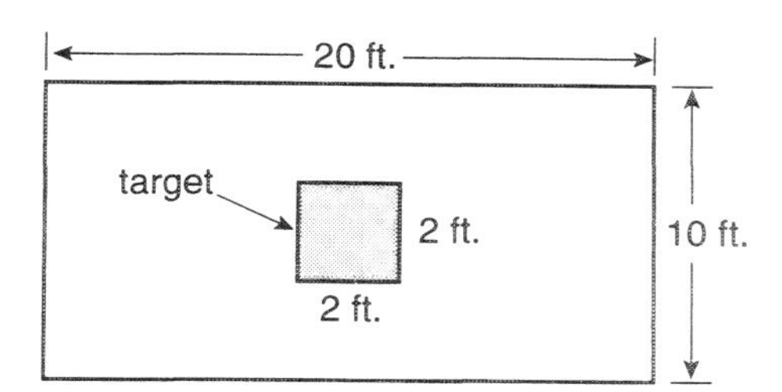

Exercise-16: The Barn's Side

17. A roulette wheel has a total of 38 positions that are numbered "0", "00", and 1–36. The positions "0" and "00" are painted green, while half of the numbers from 1–36 are painted red and the other half are painted black. A small ball is randomly spun on the wheel, and it eventually stops at one of the 38 positions. Calculate the probability that the ball:

(a) lands on a green number.
(b) lands on a red number.
(c) lands on the number 35.
(d) lands on a number between 1 and 12 inclusive.
(e) does not land on a green number.
(f) does not land on 35.

18 ADDING PROBALITIES

Since any two given events are subsets of some sample space, it makes sense to ask what the meaning of their union or intersection might be. As you will see in this lesson, the answer to this simple question will help us to solve even more useful types of problems that involve probability. Take a look at our first definition:

Definition-1: (Union & Intersection of Events)

If A and B are any events (i.e., they are both subsets of the sample space for some experiment), then:

(1) the union: $A \cup B$ represents the event that *either* A occurs, or B occurs, or *both* A and B occur.

(2) the intersection: $A \cap B$ represents the event that *both* A and B occur at the same time.

Example-1: A bin contains 2 red balls, 3 white balls, and 1 yellow ball. A ball is selected at random from the bin. The event A consists of selecting a red ball, while the event B consists of selecting a white ball. Determine the following:

(a) The sample space S.
(b) $A \cup B$
(c) The meaning of $A \cup B$

Solution:
(a) Here, we can represent the sample space S by the set:

$$S = \{R_1, R_2, W_1, W_2, W_3, Y\}$$

where subscripts have been used to let us tell the difference between the different red and white balls. Since A consists of selecting a red ball, while B is the event which consists of selecting a white ball, we have:

$$A = \{R_1, R_2\}$$
$$B = \{W_1, W_2, W_3\}$$

(b) Using the definition of the union of two sets, we now calculate $A \cup B$:

$$A \cup B = \{R_1, R_2, W_1, W_2, W_3\}$$

(c) From the result in (b), we see that $A \cup B$ can be described as the event which consists of either selecting a red ball or a white ball. This is in agreement with Definition-1.

Example-2: An experiment consists of flipping a single coin twice, and observing whether you get a heads or a tails on each toss. Let A be the event which consists of observing heads on the first toss, and let B be the event consisting of observing a heads on the second toss. Determine the following:

(a) The sample space S.
(b) $A \cup B$ and what it means.
(c) $A \cap B$ and its meaning.

Solution:
(a) The sample space S for this experiment may be written as:

$$S = \{(H,H), (H,T), (T,H), (T,T)\}$$

where the ordered pairs are used to record what happened on each toss. For example, the *ordered* pair (T,H) means that we observe a tails on the first toss, and a heads on the second toss, and so on.
(b) Since A is the event which consists of observing heads on the first toss, then we have:

$$A = \{(H,H), (H,T)\}$$

Similarly, since B consists of observing heads on the second toss, we have:

$$B = \{(H,H), (T,H)\}$$

so that:

$$\begin{aligned} A \cup B &= \{(H,H), (H,T)\} \cup \{(H,H), (T,H)\} \\ &= \{(H,H), (H,T), (T,H)\} \end{aligned}$$

and the event $A \cup B$ is seen to represent the event consisting of either observing heads on the first toss, second toss, or on both tosses.
(c) Using the information from (b), we have:

$$\begin{aligned} A \cap B &= \{(H,H), (H,T)\} \cap \{(H,H), (T,H)\} \\ &= \{(H,H)\} \end{aligned}$$

In other words, $A \cap B$ is the event consisting of observing heads on *both* tosses. Notice that this agrees with Definition-1 above.

Example-3: A single card is selected from a standard deck of 52 cards. Let A be the event which consists of selecting a red card, and let B be represent the event consisting of selecting a face card. Describe the events:

(a) $A \cup B$
(b) $A \cap B$

Solution:
(a) Using Definition-1, the event $A \cup B$ consists of *either* selecting a red card, or a face card, or both.
(b) Similarly, $A \cap B$ consists of selecting a card which is *both* a red card and a face card.
Now that we have seen some simple examples of these ideas, we are in a position to see how they are related. This relationship is given in the **Probability Addition Formula**:

Theorem-1: (Probability Addition Formula)

Let A and B be events in a finite sample space S, then:

$$P(A \cup B) = P(A) + P(B) - P(A \cap B) \qquad (1)$$

Example-4: A single die is tossed, and the number which appears on its top is observed. Use Theorem-1 to determine the following probabilities:

(a) Observing an even number or a number less than 4.
(b) Observing an odd number or a number greater than 5.

Solution:
Here, the sample space S is given by:

$$S = \{1,2,3,4,5,6\}$$

(a) Let A be the event which consists of observing an even number, and B consist of observing a number less than 4. By Theorem-1:

$$\begin{aligned} P(A \cup B) &= P(A) + P(B) - P(A \cap B) \\ &= \frac{3}{6} + \frac{3}{6} - \frac{1}{6} \\ &= \frac{5}{6} \end{aligned}$$

So the probability of observing either an even number or a number less than 4 (or both) is 5/6.

(b) Let A be the event which consists of observing an odd number, and B consist of observing a number greater than 5.

By Theorem-1:

$$P(A \cup B) = P(A) + P(B) - P(A \cap B)$$
$$= \frac{3}{6} + \frac{1}{6} - \frac{0}{6} = \frac{4}{6}$$
$$= \frac{2}{3}$$

So the probability of observing either an odd number or a number greater than 5 is 2/3.

If it is impossible for two given events to occur at the same time, then we say that these events are **mutually exclusive.** Under such circumstances, as you will see, the Probability Addition Formula becomes particularly easy to use. Let's continue with the last definition of this lesson:

Definition-2: (Mutually Exclusive Events)

The events A and B are said to be **mutually exclusive** if and only if:

$$A \cap B = \emptyset \qquad (2)$$

In other words, *two events are mutually exclusive if they cannot both occur at the same time.*

Corollary-1: (Mutually Exclusive Events)

Let A and B be mutually exclusive events in a finite sample space S, then:

$$P(A \cup B) = P(A) + P(B) \qquad (3)$$

Example-5: A card is selected at random from a standard deck of 52 cards. Determine the following probabilities:

(a) Selecting a face card or a 5.
(b) Selecting a jack or an ace.
(c) Selecting a club or a diamond.

Solution:
First, observe that each pair of events in (a)-(c) is mutually exclusive, i.e., you can't select a face card and a 5 at the same time; you can't select a jack and an ace at the same time, and so on. This means that Equation (3) can be used.
(a) Here, using less formal notation, we have:

$$\begin{aligned} P(\text{face card or a 5}) &= P(\text{face card}) + P(5) \\ &= \frac{12}{52} + \frac{4}{52} = \frac{16}{52} \\ &= \frac{4}{13} \end{aligned}$$

(b) Similarly, using Equation (3) again, we get:

$$\begin{aligned} P(\text{jack or ace}) &= P(\text{jack}) + P(\text{ace}) \\ &= \frac{4}{52} + \frac{4}{52} = \frac{8}{52} \\ &= \frac{2}{13} \end{aligned}$$

(c) Finally, we have:

$$\begin{aligned} P(\text{club or diamond}) &= P(\text{club}) + P(\text{diamond}) \\ &= \frac{13}{52} + \frac{13}{52} = \frac{26}{52} \\ &= \frac{1}{2} \end{aligned}$$

EXERCISE SET #18

PART A

In Exercises 1–10, a single die is tossed once, and the number on its top is observed. Determine the probabily of observing:

1. An even number or an odd number.
2. A number less than 2 or a number greater than 3.
3. An even number or a number less than 4.
4. An odd number or a number less than 2.
5. A 2 or a 5.
6. An even number or a 3.
7. An odd number or a 4.
8. A number less than 4 or greater than 2.
9. A number greater than 3 or less than 5.
10. An even number or a number less than 5.

In Exercises 11–20, a single card is randomly selected from a standard deck of 52 cards. Determine the probability of observing:

11. A jack or a 3.
12. An ace or a 2.
13. A queen or a red card.
14. A heart or a club.
15. A heart or a diamond.
16. An ace or a heart.
17. A jack or a club.
18. A 10 or a face card.
19. A 10 or a jack.
20. An ace or a king.

PART B

21. A shipment of fruit contains 20 oranges, 70 apples, and 30 peaches. A single piece of fruit is selected at random from the shipment. Determine the following probabilities:

    (a) P(orange or apple)
    (b) P(apple or peach)
    (c) P(orange or peach)

22. A restaurant surveyed 120 of its customers and found that:

    30 customers liked to use real butter.
    40 customers liked to use margarine.
    10 customers liked to use both butter and margarine.

    If the survey results are representative of the restaurant's customers, then determine the following probabilities:

    (a) P(customer likes butter)
    (b) P(customer likes margarine)
    (c) P(customer likes butter and margarine)
    (d) P(customer likes butter or margarine)

23. The manager of a hotel surveyed its patrons to determine their level of satisfaction. A total of 100 customers were surveyed, and it was determined that:

    40 customers liked the service.
    50 customers liked the room.
    20 customers liked both the service and the room.

    Assuming that this survey was representative of the hotel's customers, then determine the following probabilities:

    (a) P(customer likes service)
    (b) P(customer likes room)
    (c) P(customer likes both service and room)
    (d) P(customer likes room or service)

    Hint: To answer (d), you can draw a Venn Diagram, or simply use Equation (1) – this is easier!

24. Provide a proof to Corollary-1. [**Hint:** Recall that $P(\emptyset) = 0$.]
25. If A,B, and C are events in a finite sample space, prove that:

$$P(A \cup B \cup C) = P(A) + P(B) + P(C) - P(A \cap B) - P(A \cap C) - P(B \cap C) - P(A \cap B \cap C)$$

    [**Hint:** Draw a Venn diagram.]

26. If A,B, and C are events in a finite sample space such that any *pair* of *different* events is mutually exclusive, then show that:

$$P(A \cup B \cup C) = P(A) + P(B) + P(C)$$

    [**Hint:** Use Exercise-25.]

19 CONDITIONAL PROBABILITY

In some cases, the probability that a given event occurs may depend upon whether one or more related events have taken place. In order to see why this might happen, let's consider the experiment which consists of flipping a single die. We know that the sample space for this experiment is:

ROLLING A SINGLE DIE

$$S = \{1,2,3,4,5,6\}$$

If we were to ask the question: "What is the probability of observing a 2?", then the answer is simply 1/6. Now suppose, however, that we roll the die, and someone tells us that an even number was observed. If we now *know* that an even number was observed, then the probability of observing a 2 is now 1/3 since there are only three even numbers in the sample space S. We call this latter case a **conditional probability** since the the probability of observing a 2 is dependent upon the *condition* that an even number was observed. In shorthand notation, we express this probability as:

$$P(\text{observing a 2} \mid \text{even number}) = \frac{1}{3}$$

which is read "the probability of observing a 2 *given* the occurrence of an even number." Our first definition provides us with a method of calculating conditional probabilities in general.

Definition-1: (Conditional Probability)

The **conditional probability** of any event A *given* that the event B has occured, is given by:

$$P(A \mid B) = \frac{P(A \cap B)}{P(B)} \quad \text{(provided } P(B) \neq 0\text{)} \qquad (1)$$

where the symbol $P(A \mid B)$ is read "the probability of A given B" or "the probability of A given the occurrence of B."

Example-1: A single card is selected from a deck of 52 playing cards. Find the probability that it is a heart, given that the card is red.

Solution:
(a) If we take a common sense approach to this problem, then we know that there are 26 red cards in a standard deck, and exactly half of these are hearts. So we would conjecture that:

$$P(\text{heart} \mid \text{red card}) = \frac{1}{2}$$

But let's solve this problem by using Equation (1), and see if our common sense approach is consistent with Definition-1. According to (1),

$$P(\text{heart} \mid \text{red card}) = \frac{P(\text{heart and red card})}{P(\text{red card})}$$

(b) Since there are 13 hearts in the deck, and all of them are red, then:

$$P(\text{heart and red card}) = \frac{13}{52} = \frac{1}{4}$$

while there are 26 red cards in the deck, so:

$$P(\text{red card}) = \frac{26}{52} = \frac{1}{2}$$

(c) Substituting the probabilities we obtained in (b) into (1) we get:

$$P(\text{heart} \mid \text{red card}) = \frac{P(\text{heart and red card})}{P(\text{red card})} = \frac{1/4}{1/2} = \frac{1}{2}$$

and this result agrees with the one that we obtained by just using our mathematical intuition.

Example-2: A total of 120 customers at Joe's Steakhouse were surveyed to determine the degree of customer satisfaction with the quality of their food. The results of this survey appear below:

Survey Results for Joe's Steakhouse			
Sex of Customer	**High Quality**	**Low Quality**	**Totals**
Males	45	25	70
Females	40	10	50
Totals:	85	35	120

Find the probability that:
(a) a customer feels the food is of high quality, given that the customer is male.
(b) a customer is female, given that the customer feels the food is of low quality.

Solution:
(a) Let A be the event "the food is of high quality", and let B represent the event "the custumer is male." Then by Equation (1) we have:

$$P(A \mid B) = \frac{P(A \cap B)}{P(B)} = \frac{45/120}{70/120} = \frac{9}{14}$$

Note here that $P(A \cap B) = 45/120$ since the above table shows that there are 45 out of the 120 people surveyed who are male *and* who felt that the food was of high quality.
(b) Let A be "the customer is female" and B be "the food is of low quality." Once again, according to Equation (1):

$$P(A \mid B) = \frac{P(A \cap B)}{P(B)} = \frac{10/120}{35/120} = \frac{2}{7}$$

Observe that we set $P(A \cap B) = 10/120$ since there were 10 people out of the 120 people surveyed who were both female *and* who thought the food was of low quality.

Example-3: A single coin is flipped twice. Given that heads appeared on the first toss, what is the probability that tails will appear on the second toss?

Solution:
(a) From previous work, we know that the sample space S for this experiment is given by:

$$S = \{(H,H), (H,T), (T,H), (T,T)\}$$

Let A be the event of observing tails on the second toss, and B consist of observing heads on the first toss. Then:

$$A = \{(H,T), (T,T)\} \text{ and } B = \{(H,H), (H,T)\} \text{ so } A \cap B = \{(H,T)\}$$

Consequently, $P(A \cap B) = 1/4$ and $P(B) = 1/2$.

(b) Substituting this information into Equation (1), we get:

$$P(A \mid B) = \frac{P(A \cap B)}{P(B)} = \frac{1/4}{1/2} = \frac{1}{2}$$

In the last example, observe that the probability of getting tails on the second toss was not really affected by the fact that heads occured on the first toss since if this fact were not given, then the probability of observing a tails on any given toss of the coin is $1/2$ anyway.

Thus, we say that these events are *independent*. We now formalize this idea in the next definition:

Definition-2: (Independent Events)

The events A and B are said to be **independent** if:

$$P(A \mid B) = P(A) \text{ or } P(B \mid A) = P(B) \tag{2}$$

Thus, if two events are independent, then the occurrence of any one event does not affect the probability of occurrence of the other event.

Theorem-1: (Independent Events)

If A and B are independent events, then:

$$P(A \cap B) = P(A) \cdot P(B) \tag{3}$$

Example-4: A single coin is flipped twice. Determine the probability of obtaining heads on both tosses.

Solution:
Since the event which consists of getting heads on the second toss is independent of getting heads on the first toss [A coin doesn't have a memory!], then by Theorem-1:

$$\begin{aligned} P(\text{heads and heads}) &= P(\text{heads}) \cdot P(\text{heads}) \\ &= \frac{1}{2} \cdot \frac{1}{2} \\ &= \frac{1}{4} \end{aligned}$$

Example-5: A sharpshooter hits a target 90% of the time. If the sharpshooter fires at the target twice, then what is the probability that he misses both times? Assume that the events are independent.

Solution:
(a) The probability that the sharpshooter hits the target must be $P(\text{hit}) = 9/10$. Consequently, the probability that the sharpshooter misses the target is:

$$\begin{aligned} P(\text{miss}) &= 1 - P(\text{hit}) = 1 - \frac{9}{10} \\ &= \frac{1}{10} \end{aligned}$$

(b) By Theorem-1, we then have:

$$\begin{aligned} P(\text{miss and miss}) &= P(\text{miss}) \cdot P(\text{miss}) \\ &= \frac{1}{10} \cdot \frac{1}{10} \\ &= \frac{1}{100} \end{aligned}$$

Exercise Set #19

PART A

In Exercises 1–10, a single die is tossed once, and the number on its top is observed. Determine the conditional probabily of observing:

1. a 4, given that the number is even.
2. a 2, given that the number is even.
3. a 3, given that the number is odd.
4. a 5, given that the number is odd.
5. a 3, given that the number is less than 5.
6. a 4, given that the number is less than 5.
7. a 6, given that the number is greater than 4.
8. a 5, given that the number is greater than 4.
9. a 6, given that the number is evenly divisible by 3.
10. a 2, given that the number is less than 3.

In Exercises 11–20, a single card is drawn from a standard deck of 52 playing cards. Find the conditional probability of observing:

11. a diamond, given that the card is red.
12. a 3 of hearts, given that the card is red.
13. a 2 of diamonds, given that the card is red.
14. a 3 of diamonds, given that the card is a diamond.
15. an ace of clubs, given that the card is a club.
16. an ace of clubs, given that the card is an ace.
17. a face card, given that the card is red.
18. an ace, given that the card is a face card.
19. a queen, given that the card is a face card.
20. a 2, given that the card selected is not a face card.

PART B

21. A certain family has a total of two children. Assume that the probability of having a girl is the same as the probability of having a boy.

    (a) Show that the sample space for this experiment can be written:

    $$S = \{(B,B), (B,G), (G,B), (G,G)\}$$

    where ordered pairs have been used to denote the order in which the children are born.

(b) What is the conditional probability that the second child is a boy given that the first child is a girl?
(c) What is the conditional probability that the second child is a boy given that the first child is a boy?
(d) What is the conditional probability that the second child is a girl given that the first child is a boy?

In Exercises 22–28, a pair of dice are rolled. Determine the conditional probability that the top numbers on the dice add up to:

22. 7, given that the first die is 2.
23. 7, given that the first die is 5.
24. 6, given that the first die is 4.
25. 6, given that the first die is 3.
26. a number greater than 10, given the first die is 5.
27. a number less than 4, given the first die is 2.
28. an even number, given that the first die is an odd number.
29. A total of 160 guests at the Crown Hotel were surveyed to determine their satisfaction with room accommodations. The survey results were as follows:

GUEST SURVEY AT CROWN HOTEL			
Guest Type	**Good Rating**	**Poor Rating**	**Totals:**
Business	70	10	80
Non-Business	50	30	80
Totals:	120	40	160

Determine the following probabilities:

(a) A guest feels the accommodations are good, given that the guest is a non-business guest.
(b) A guest feels the accommodations are poor, given that the guest is a business guest.
(c) A guest feels that the accommodations are good.
(d) A guest is a business guest, given that the guest feels the accommodations were poor.

30. According to records at the Baseball Hall of Fame, Mickey Mantle, who played for the New York Yankees for 18 years had the following statistics:

PLAYER STATISTICS FOR MICKEY MANTLE	
At Bats	8,102
Hits	2,415
Doubles	344
Triples	72
Home Runs	536
Walks	1,734
Strike Outs	1,710

Determine the following conditional probabilities:

(a) the probability of hitting a home run, given that he got a hit.
(b) the probability of striking out given that he did not get a hit.
(c) the probability of hitting a triple, given that he got a hit.
(d) the probability of walking, given that he did not get a hit.

31. A single die is tossed two times, and the number on top is observed each time. Determine each of the following probabilities:

    (a) observing a 2 both times.
    (b) observing an even number both times.
    (c) observing an odd number each time.

    Hint: The outcome of the second toss is independent of the first toss. A die doesn't have a memory!

32. Roughly, 2% of a shipment of refrigerators is known to be defective. If two refrigerators are selected at random from the shipment, what is the probability that both of them are defective? Assume independence.
33. A sharpshooter hits the target 95% of the time. If we assume the independence of events, and he fires at the target twice, what is the probability that:

    (a) he hits the target both times.
    (b) he misses the target both times.
    (c) he hits the target at least once.

34. A certain pain reliever is known to help patients suffering from severe headaches 30% of the time. If two headache sufferers are treated, then what is the probability that both patients will be helped by the medication? Assume independence.

35. The concept of independent events may be generalized as follows: we say that the events $E_1, E_2, \ldots, E_n$ are independent if and only if:

$$P(E_1 \cap E_2 \cap \ldots \cap E_n) = P(E_1) \cdot P(E_2) \cdots P(E_n) \qquad (4)$$

Use Equation (4) to solve Exercise-33 assuming that the sharpshooter now fires a total of *four times* at the target.

36. Show that if an event A is independent of an event B so that:

$$P(A \mid B) = P(A)$$

then we can also conclude that:

$$P(B \mid A) = P(B)$$

37. Show that if a fair coin is tossed n-times, then the probability of obtaining at least one head is given by:

$$P(\text{at least on head}) = 1 - \frac{1}{2^n}$$

What happens to the magnitude of this probability as the number of tosses n increases?

38. An unusual dart game is constructed in the xy-plane by inscribing a quarter-circle inside the square as shown below. A blindfolded student then randomly throws a dart at the square. What is the probability that a dart falls inside the quarter-circle *given* that it falls inside the square?

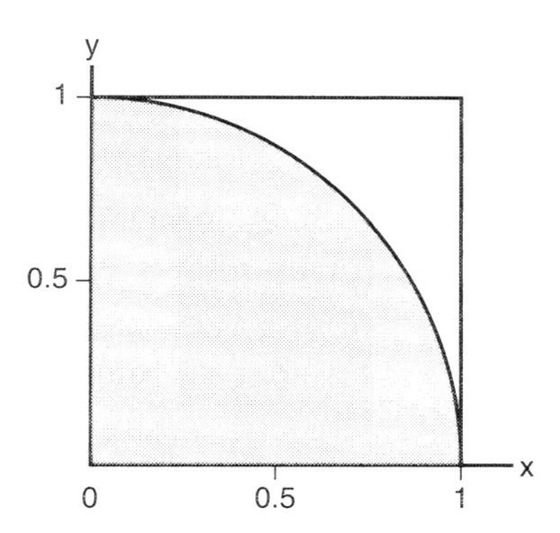

Exercise-38: An Unusual Dart Game

20 EXPECTED VALUE

A **random variable** is a rule which assigns a real number to each element of a sample space S. To avoid confusion, we'll usually denote random variables by using capital letters such as X or Y. While this concept may appear to be quite abstract at first glance, the next two examples should clarify its meaning.

Example-1: An experiment consists of tossing two coins at the same time, and observing the number of heads which appear. We know that the sample space S for this experiment can be written as:

$$S = \{(T,T), (T,H), (H,T), (H,H)\}$$

If we let X denote the number of heads observed, then we can assign a real number to each element of the sample space as follows:

EXAMPLE-1: FLIPPING TWO COINS	
Outcome of Experiment	**X (#Heads Observed)**
(T,T)	0
(T,H)	1
(H,T)	1
(H,H)	2

From the table, we see that we have assigned a real number, i.e., the number of heads observed to each element of the sample space S, so we see that X is a random variable.

Example-2: An experiment consists of firing a bullet at a circular target 100 times. If we let Y denote the distance between the center of the target and the point of impact, we see that each outcome is assigned a real number. Once again, Y is a random variable.

In the first example, observe that the random variable X can only assume a *finite number* of different values, i.e.,

$$X = 0,1,2$$

so we say that X is a **discrete random variable,** while in the second example, Y can assume *any real value* within an entire interval of real numbers, and consequently, we call Y a **continuous random variable.**

Definition-1: (Expected Value)

Let X be a discrete random variable which assumes the n-possible values:

$$x_1, x_2, x_3, \ldots, x_n$$

whose corresponding probabilities of occurrence are:

$$P(x_1), P(x_2), P(x_3), \ldots, P(x_n)$$

then the **expected value** of X, denoted by $E(X)$, is given by:

$$E(X) = P(x_1)x_1 + P(x_2)x_2 + P(x_3)x_3 + \ldots + P(x_n)x_n \qquad (1)$$

In other words, *to calculate the expected value of a discrete random variable, we multiply each possible value of the random variable by its corresponding probability, and add all of the resulting products.*

Example-3: An experiment consists of tossing a single die and observing the number which appears on top. If X denotes the number which appears, then find its expected value.

Solution:

(a) Here, there are only 6 possible values for X; namely,

$$X = 1,2,3, \ldots, 6$$

and since each outcome is equally likely, each possible value of X must have the same probability of 1/6. These possible values of X and their corresponding probabilities have been listed below:

TOSSING A SINGLE DIE

Possible Values of X	Corresponding Probabilities
1	1/6
2	1/6
3	1/6
4	1/6
5	1/6
6	1/6

(b) According to Definition-1, we now calculate the expected value of X by multiplying each possible value of X and its corresponding probability, and adding up all such products:

$$\begin{aligned}E(X) &= (1\cdot\frac{1}{6})+(2\cdot\frac{1}{6})+(3\cdot\frac{1}{6})+(4\cdot\frac{1}{6})+(5\cdot\frac{1}{6})+(6\cdot\frac{1}{6})\\ &= \frac{1}{6}+\frac{2}{6}+\frac{3}{6}+\frac{4}{6}+\frac{5}{6}+\frac{6}{6}\\ &= \frac{7}{2}\\ &= 3.5\end{aligned}$$

Notice that our answer, 3.5, is the **average value** of X which we would expect if we were to repeat the experiment many times.

Example-4: In a fundraiser for the needy, 200 raffle tickets are sold at $1.00 each. Only one $100 prize is to be awarded to the winner. John buys one raffle ticket. Find the expected value of John's *net* winnings.

Solution:
(a) If we let X represent the *net* amount that John wins, then X has only two possible values:

$$x_1 = \$99 \text{ and } x_2 = -\$1$$

since if he wins, his net winnings are $100 − $1 (less the price of the ticket), and if he loses, then he loses just $1. Observe that a minus sign has been used to denote a loss.
(b) Since there are 200 raffle tickets, then the probability that he wins is simply 1/200 while the probability that he loses is just 1 − 1/200 or 199/200. In other words,

$$P(x_1) = \frac{1}{200} \text{ and } P(x_2) = \frac{199}{200}$$

(c) Using the definition of expected value, we get:

$$\begin{aligned}E(X) &= P(x_1)x_1 + P(x_2)x_2\\ &= \frac{1}{200}(\$99) + \frac{199}{200}(-\$1)\\ &= \frac{\$99}{200} - \frac{\$199}{200} = -\$\frac{100}{200}\\ &= -\$0.50\end{aligned}$$

Thus, in the long run, John would *lose* an average of 50 cents on each lottery ticket purchased.

In general, we say any game of chance is a **fair game** if the expected value of our net winnings is zero. This means that in the long run, we would expect to win as much as a we lose. It is clear that gambling casinos make a considerable amount of money because the games that they offer are not fair, and more importantly, their corresponding expected values (like the previous example) are negative.

Take a look at the last example for this lesson.

Example-5: Assuming the same information as in Example-4, what *price* should be charged for each raffle ticket, if the raffle is to be a fair game?

Solution:

(a) Let p be the unknown price. In order for the game to be fair, the expected value of John's net winnings must be zero. If we let X represent the net amount that John wins, then X has only two possible dollar values:

$$x_1 = (100 - p) \text{ and } x_2 = -p$$

since if he wins, his net winnings are $100 - p$, and if he loses, then he loses just p. Once again, a minus sign has been used to denote a possible loss.

(b) As in Example-4, we have:

$$P(x_1) = \frac{1}{200} \text{ and } P(x_2) = \frac{199}{200}$$

(c) So by (a) and (b), we want:

$$E(X) = P(x_1)x_1 + P(x_2)x_2 = 0$$

or,

$$\frac{1}{200}[(100 - p)] + \frac{199}{200}[(-p)] = 0$$

If we now multiply both sides of this equation by 200, we get:

$$(100 - p) - 199p = 0$$

$$100 - 200p = 0$$

$$p = \frac{1}{2} = 0.50$$

So our solution is p = \$0.50. Thus, in order to be a fair game, we should charge \$0.50 per raffle ticket. We call this price a **fair price,** since it results in an expected value of the net winnings which is equal to zero-it is fair to both the player and the "house."

EXERCISE SET #20

PART A

1. A certain "wheel of fortune" has the numbers 1–10 equally spaced along its periphery. Let the random variable X denote the number which comes out when the wheel is spun. Find the expected value of X.
2. A tetrahedral die [A tetrahedron is a regular solid with four sides.] has its faces numbered 1 thru 4. Let X denote the number which is observed when the die is tossed. Find $E(X)$, the expected value of X.
3. A totally unprepared student is asked to take a 100 question multiple-choice exam in Ancient History. Each question has four possible answers, and no question can be left blank. Let X represent the total number of correct answers on the student's answer sheet. If the student answers each question at random, then find the expected value $E(X)$ of the number of correct questions.
4. A game of chance consists of rolling a single die. If a number greater than 4 appears on its top, then you win; otherwise you lose. If it costs you $1 each time you play, and you receive $2 each time you win, then determine the following:

   (a) The expected value of your winnings.
   (b) What price should be charged to make this a fair game.

5. A church sells 400 raffle tickets for $2 each, and a single prize of $100 is to be awarded to the winner.

   (a) If Bob buys one raffle ticket, what is the expected value of his winnings?
   (b) How much money would you *expect* the church to make on this raffle?
   (c) Find the fair price of a raffle ticket.

6. A game consists of tossing a pair of dice and observing the two numbers which appear on top. If both numbers are the same, then you win $3; otherwise you lose. It costs you $1 each time you play, and you receive $3 each time you win.

   (a) Find the expected value of your winnings.
   (b) Determine the fair price for this game.
   (c) If you played this game 100 times, then how much would you *expect* to lose?

7. A roulette wheel consists of the numbers 0, 00, and 1–36 spaced evenly along its edge. A small ball is spun, and falls at random into one of the numbered slots on the wheel. If you bet on a single number, it costs you $5; but if you win, then the house pays out $175.

(a) Determine the expected value of your winnings.
(b) Determine a fair price for this game.
(c) If you played game 10 times, then how much would you expect to lose?

8. A sharpshooter is known to hit a target 90% of the time. If he misses the target, then you win \$10. If he hits the target, then you must pay him \$2.

   (a) Find the expected value of your winnings.
   (b) Determine a fair price for this game.

9. A certain card game consists of selecting one card from a standard deck of playing cards. If you select an ace, then you win \$3; if you select a king then you win \$2. If any other card appears, you lose \$1.

   (a) Determine the expected value of your winnings.
   (b) Find the fair price for this game.

PART B

10. Let X be a random variable which represents the sum of the numbers which appear on the tops of a pair of dice when they are tossed.

    (a) Show that the entries in the following table are valid:

TOSSING A PAIR OF DICE

Sum of Numbers X	Probability of Sum $P(X)$
2	1/36
3	2/36
4	3/36
5	4/36
6	5/36
7	6/36
8	5/36
9	4/36
10	3/36
11	2/36
12	1/36

(b) Show that: $P(2) + P(3) + P(4) + \ldots + P(12) = 1$.
(c) Find the expected value $E(X)$.

11. A jar contains a total of N raffle tickets that are numbered consecutively 1,2,3,. . .,N. Let X denote the number of the raffle ticket which is randomly selected. Show that:

$$E(X) = \frac{N+1}{2}$$

Hint: It is a well-known fact that:

$$1+2+3+\ldots+N = \frac{N(N+1)}{2}$$

12. A non-profit organization sells N raffle tickets at a cost of C dollars each. Bob purchases one ticket, and will receive a prize of W dollars if he wins.
(a) Show that the expected value of Bob's winnings is:

$$E(X) = \frac{W - CN}{N}$$

(b) Show that the fair price P for this game is given by:

$$P = \frac{W}{N}$$

(c) From (a) and (b), show that:

$$E(X) + C = P$$

In other words,

Expected Value + Cost to Play = Fair Price

13. A certain die is not fair and the probability that its various sides appear on top is given below:

A Lopsided Die	
Number on Face, X	Probability $P(X)$
1	0.10
2	0.20
3	0.25
4	0.15
5	0.20
6	0.10

Find the expected value $E(X)$ of the number on the face.

14. In a certain game of chance, Bob plays with the lopsided die in Exercise-13. If it costs him $1 to play, and he wins $3 each time a number greater than 4 comes out, then find the expected value of Bob'swinnings.
15. The position of bug sitting on the real axis is given by the random variable X. The bug can step only one unit to the right or one unit to the left each time it moves. The probability that the bug moves to left or to the right are 0.45, and 0.55 respectively. Assuming the bug starts out at the origin ($X = 0$), and then takes 100 steps, what is the expected value of the bug's position?

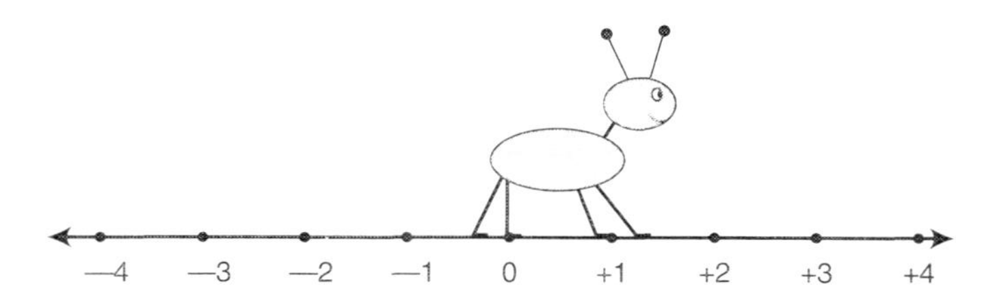

Exercise-15: A Bug Crawling on the Real Axis

31. $10 \div (1 + i)$
32. $(2 + i) \div (2 + 2i)$
33. $(4 - 2i) \div (2 - i)$
34. $(1 + i) \div (2i)$
35. $(2i) \div (3 + 2i)$

In Exercises 36–40, simplify each of the given expressions. Be sure to write your answers in the form $a + bi$.

36 $$\frac{(1+i)^2(2+i)}{(1-i)}$$

37 $$\frac{(1+2i)(1-2i)}{5i}$$

38 $$\frac{[(1+3i)-(2-2i)]}{(2+4i)}$$

39 $$\frac{(2-3i)\cdot[(1+2i)+(-1+2i)]}{(3-4i)}$$

40 $$\frac{(2+2i)(2-2i)(1-i)(4+4i)}{16i}$$

PART B

41. Prove Theorem-1.
42. Derive Equation (5), i.e., the equation for complex division.
43. Show that for any complex numbers z_1 and z_2:

$$\overline{(z_1 + z_2)} = \bar{z}_1 + \bar{z}_2$$

44. If z is a complex number such that $z = \bar{z}$, then what can you conclude about z?
45. Show that for any complex numbers z_1 and z_2:

$$\overline{(z_1 \cdot z_2)} = \bar{z}_1 \cdot \bar{z}_2$$

46. Let $z = a + bi$ be any complex solution of the n-th degree polynomial equation:

$$a_n z^n + a_{n-1} z^{n-1} + a_{n-2} z^{n-2} + \ldots. + a_1 z + a_0 = 0$$

where $a_n, a_{n-1}, \ldots, a_0$ are real numbers, and $a_n \neq 0$. Show that $\bar{z} = a - bi$ is also a solution of this equation.

21 Simple Interest and Simple Discounts

In the next three lessons, we shall provide an introduction to what is commonly called **Consumer Mathematics,** or the **Mathematics of Finance.** In this lesson, we begin our financial journey with a basic discussion of simple interest.

Interest is nothing more than the price that is normally paid for the use of money. Simple interest is the easiest type of interest to calculate and its definition is given below:

Definition-1: (Simple Interest)

If the **principal** P is an amount of money which is borrowed or invested at an **annual interest rate** r (decimal form) for a period of t years, then the **simple interest** I is defined to be:

SIMPLE INTEREST FORMULA

$$I = Prt \qquad (1)$$

Example-1: A student took out a loan of \$500 for two years, at a simple interest rate of 5% per year. What is the interest charge that the student must pay?

Solution:

(a) Here, our principal is \$500, our annual interest rate is 0.05 since 5% = 0.05 when expressed in decimal form, and the period or duration of the loan is $t = 2$ years. According to Definition-1, the interest charge that must be repaid is:

$$\begin{aligned} I &= Prt \\ &= (\$500)(.05)(2) \\ &= \$50.00 \end{aligned}$$

Now, as we mentioned in Definition-1, money can be either *borrowed* or *invested* using the simple interest model. In order to make things definite, suppose that a principal P is borrowed over a period of t years at a simple interest rate of r. A question which arises is how do we determine the total amount of money that you owe at the end of t years?

If we let A be the total amount of money owed, then it is clear that we must have:

$$\text{Total Amount Owed} = \text{Principal} + \text{Interest}$$

or symbolically,

$$A = P + I \tag{2}$$

If we now substitute Equation (1) into (2), we obtain:

$$\begin{aligned} A &= P + I \\ &= P + Prt \\ &= P(1 + rt) \end{aligned}$$

so that a useful formula for the total amount owed is:

TOTAL AMOUNT OWED

$$A = P(1 + rt) \tag{3}$$

In Equation (3), we sometimes refer to P as the **present value,** and to A as the **future value.**

It is important to realize that (3) still works even when you invest money at some simple interest rate r over a period of t years. In this case, A represents the amount of money which is *owed to you,* not the amount of money that you owe. Take a look at the next two examples.

Example-2: A small loan company lent a total of $12,000 to a business owner for 9 months at an interest rate of 8% per year. How much money must the business owner pay back to the loan company at the end of the 9 month period?

Solution:
(a) Here, we can make the following identifications:

$$\begin{aligned} P &= \$12{,}000 \\ r &= 0.08 \\ t &= 0.75 \text{ years} \end{aligned}$$

Consequently, by (3) we obtain the total amount A as:

$$\begin{aligned} A &= P(1 + rt) \\ &= (\$12{,}000)[1 + (0.08)(0.75)] \\ &= \$12{,}000 \cdot (1.06) \\ &= \$12{,}720 \end{aligned}$$

Thus, at the end of 9 months, the business owner must pay back a total amount of $12,720.

Example-3: A student borrowed $1200 for a period of 6 months. The total amount that the student had to pay back after 6 months was $1,272. What was the simple interest rate charged?

Solution:
(a) We can make the following identifications:

$$\begin{aligned} A &= \$1{,}236 \\ P &= \$1{,}200 \\ t &= 0.5 \text{ years} \\ r &= \text{interest rate (unknown)} \end{aligned}$$

Using Equation (3), we obtain:

$$\$1236 = \$1200 \cdot [1 + r(0.5)]$$

(b) We now solve for r to get:

$$\begin{aligned} r(0.5) &= \frac{\$1236}{\$1200} - 1 = \frac{72}{1200} \\ r &= 0.06 \end{aligned}$$

Thus, we see that the simple (annual) interest rate charged was 6%.

In some instances, a loan may be transacted by using what is called a **discount note.** In such cases, a lender deducts the interest that must be paid *at the time* a given loan is made. The total amount of interest deducted or discounted from the loan is called a **simple discount.**

On the other hand, the *net* amount of money that the borrower receives (after discounting) is called the **proceeds,** and the amount borrowed is called the **maturity value.** If we keep in mind that the simple discount is just the simple interest deducted when the loan is made, then we have:

SIMPLE DISCOUNT FORMULA

$$D = Mrt \tag{4}$$

where :

$$\begin{aligned} D &= \text{simple discount} \\ M &= \text{maturity value} \\ r &= \text{simple discount rate (decimal)} \\ t &= \text{time (in years)} \end{aligned}$$

Similarly, using the definition of proceeds and the maturity value, we have:

$$\text{Proceeds} = \text{Maturity Value} - \text{Simple Discount} \tag{5}$$

If we now substitute (4) into (5), then we obtain:

PROCEEDS FORMULA

$$P = M(1 - rt) \tag{6}$$

Example-4: A business owner signs a discount note and agrees to pay \$1,200 in 9 months at a simple discount rate of 8%. How much money does the business owner receive when the loan is transacted?

Solution:
(a) Here, we are trying to the determine the proceeds P. If we first make the identifications:

$$M = \text{maturity value} = \$1{,}200$$
$$r = 0.08 = \text{simple discount rate}$$
$$t = 0.75 \text{ years}$$

then by Equation (6) we have:

$$\begin{aligned} P &= M(1 - rt) \\ &= (\$1200)[1 - (.08)(0.75)] \\ &= \$1{,}128 \end{aligned}$$

Thus, the business owner receives \$1,128 "in hand" when the loan is successfully transacted.

EXERCISE SET #21

PART A

In Exercises 1–10, calculate the simple interest I given the principal P, annual interest rate r, and the time t.

1. $P = \$4{,}300$, $r = 6\%$, $t = 2$ years
2. $P = \$1{,}200$, $r = 5.5\%$, $t = 18$ months
3. $P = \$1{,}800$, $r = 4\%$, $t = 6$ months
4. $P = \$12{,}100$, $r = 3.5\%$, $t = 2$ years
5. $P = \$1{,}300$, $r = 4.25\%$, $t = 8$ months
6. $P = \$5{,}200$, $r = 5.5\%$, $t = 30$ months
7. $P = \$7{,}300$, $r = 8.25\%$, $t = 18$ months
8. $P = \$4{,}100$, $r = 3.5\%$, $t = 3$ years
9. $P = \$11{,}200$, $r = 3.6\%$, $t = 2$ years
10. $P = \$7{,}500$, $r = 8\%$, $t = 6$ months

PART B

11. A business owner borrowed a total of $12,500 from a bank for 9 months at an interest rate of 7.5% per year. How much money must be business owner pay back the bank at the end of the 9 month period?
12. A small loan company lent a total of $24,200 to a restaurant owner for renovations at the restaurant. The period of the loan was 18 months at an interest rate of 6% per year. How much money must the restaurant owner pay back to the loan company at the end of this time period?
13. A student borrowed $2,200 for a period of 8 months. The total amount that the student had to pay back after 8 months was $2,482. What was the simple interest rate charged?
14. Bob borrowed $4,300 for a period of 18 months. He had to pay back a total amount of $4,850 at the end of this 18 month period. What was the simple interest rate the Bob was charged?
15. Alice borrowed money for 3 years by giving her bank a discount note for $5,200. The discount rate used was 8%. Calculate:

    (a) the discount
    (b) the proceeds

16. Bob borrowed money for 18 months by giving the Happy Savings Bank a discount note for $3,000. The proceeds received by Bob amounted to $2,550. What simple discount rate was used by the bank?

17. A business owner signed a discount note and agreed to pay $1,800 in 9 months at a simple discount rate of 7%. How much money did the business owner receive when the loan is transacted?
18. A parent invested a total of $18,000 for 18 months to help pay for her daughter's college expenses. If the simple interest rate used was 4.5%, then what is the total interest that she will receive at the end of the 18 month period?
19. Alice gave her bank a discount note for $6,000 for 6 months. The discount rate was 8%. Use this information to find the proceeds.
20. What amount of money should you invest today, at 6% per year in order for it to yield a future value of $8,200 at the end of two years?

22 Compound Interest

In the simple interest model that we discussed in the previous lesson, the future value A of a given sum of money P that was invested at a simple interest rate r for a period of t years was given by:

$$\begin{aligned} A &= P(1 + rt) \\ &= P + Prt \end{aligned}$$

If you examine this formula carefully, you will notice that under the simple interest model, the original sum of money P increases by the *fixed* amount $P \cdot r$ from the end of one year to the next. In order to see this, observe that the change in future value ΔA over a one year period (i.e., from t to $t + 1$ years) is:

$$\begin{aligned} \Delta A &= P[1 + r(t + 1)] - P(1 + rt) \\ &= P \cdot r \end{aligned}$$

This phenomenon takes place because the simple interest rate r is always applied to the *same* principal P during each interest period or **conversion period.**

On the other hand, in the **compound interest model,** the interest due at the end of each conversion period is *added* to the *previous* principal. The interest for the next conversion period is then based upon this new value of the principal, and this **compounding** process continues until all of the conversion periods have been calculated.

In order to understand how this works, let's look at a simple example.

Example-1: A principal of $P = \$1000$ is invested at an interest rate of 8% compounded quarterly, i.e., every 3 months. How much money will we have at the end of just one year?

Solution:

(a) First, we will compute the interest earned at the end of the first quarter using the usual formula:

$$I = Prt = (\$1000)(.08)(0.25) = \$20.00$$

(b) We now add the original principal of $1000 to this interest to obtain a new principal of $1020 at the start of the second quarter. The interest earned at the end of the second quarter is then:

$$I = (\$1020)(.08)(0.25) = \$20.40$$

(c) We now add the previous principal of $1020 to the interest of $20.40 to obtain a new principal of $1040.40 at the start of the third quarter. The interest earned at the end of the third quarter is now:

$$I = (\$1040.40)(.08)(0.25) = \$20.81$$

(d) Finally, we add the previous principal of $1040.40 to the interest of $20.81 to obtain a principal of $1061.21 at the start of the fourth quarter. The interest earned at the end of the fourth quarter is now:

$$I = (\$1061.21)(.08)(0.25) = \$21.22$$

If we add this interest of $21.22 to the previous principal of $1061.21, then *we see that our initial investment of* $1000 *will grow to a future value of* $1082.43 *after just one year.*

This example clearly shows the difference between simple interest and compound interest. If we had invested the same amount of money, i.e., $1000, at a simple interest rate of 8% for one year, then its future value (principal + interest) would have been $1080. The difference of $2.43 between the compound interest and simple interest models is the result of compounding the principal at the end of each conversion period.

The previous example also demonstrates the need for an easier method for computing compound interest. This method is given in the first theorem of this lesson.

Theorem-1: (Compound Interest Formula)

If a sum of money P is invested at an annual interest rate of r (in decimal form) over a total of n conversion periods, then the accumulated amount A at the end of n such periods is:

COMPOUND INTEREST FORMULA

$$A = P(1 + i)^n \tag{1}$$

where:

A = Accumulated Amount (or Future Value)
P = Principal (or Present Value)
n = Number of conversion periods

Theorem-1 (Continued)

and where i is the **interest rate per conversion period,** and is given by:

INTEREST RATE PER CONVERSION PERIOD

$$i = \frac{r}{m} = \frac{\text{annual interest rate}}{\text{number of conversion periods per year}} \quad (2)$$

Observe that Equation (1) can be rewritten in the useful form:

PRESENT VALUE FORMULA

$$P = \frac{A}{(1+i)^n} \quad (3)$$

This formula is useful in determining the present value, i.e., how much money must be invested over a certain period of time in order to generate a given, total amount A.

Example-2: Use the compound interest formula to rework Example-1 above.

Solution:

(a) Here, we have a total of $n = 4$ conversion periods, and since there are four conversion periods per year, then we have $m = 4$ as well. Thus, the interest rate per conversion period is:

$$i = \frac{.08}{4} = .02$$

(b) Using an initial principal of $P = \$1000$, the compound interest formula gives:

$$\begin{aligned} A &= P(1 + i)^n \\ &= (\$1000)(1 + .02)^4 \\ &= (\$1000)(1.08243) \\ &= \$1082.43 \quad \textbf{Solution} \end{aligned}$$

In calculating the value of $(1.02)^4$, you should use the:

$\boxed{y^x}$ or $\boxed{x^y}$

key on your calculator. Observe that our answer in Example-2 agrees with the one we obtained in Example-1 by using the "brute force method."

Example-3: A credit union pays 6% per year, compounded semi-annually (every six months). If an amount of $1,200 is placed in a savings account and allowed to grow for three years, then how much money will there be in the account at the end of this period?

Solution:
(a) Here, we have a total of $n = 6$ conversion periods, and since there are two conversion periods per year, then we have $m = 2$. Now, the interest rate per conversion period is:

$$i = \frac{.06}{2} = .03$$

(b) Using an initial principal of $P = \$1200$, the compound interest formula gives:

$$\begin{aligned} A &= P(1 + i)^n \\ &= (\$1200)(1 + .03)^6 \\ &= (\$1200)(1.\ 1940523) \\ &= \$1432.86 \quad \textbf{Solution} \end{aligned}$$

From the previous examples, it is evident that the future value is a function of several factors; not only does it depend upon the annual interest rate, but also upon the total number of conversion periods per year. We need therefore, some simple method for comparing various annual rates. This leads us to the concept of the **effective interest rate.**

Definition-1: (Effective Interest Rate)

Suppose that a principal is invested at an annual interest rate r, which is compounded several times per year, to produce a total amount of money A after one year. Then the **effective interest rate** r_E is the *simple interest rate* that would generate the *same* total amount of money A after one year.

Theorem-2 (Effective Interest Rate)

If a principal P is invested at an annual interest rate r, which is compounded m times per year, then the effective interest rate r_E is given by:

EFFECTIVE INTEREST RATE

$$r_E = (1 + i)^m - 1 \tag{4}$$

Example-4: A certain investment pays 7% compounded semi-annually. What is the effective interest rate?

Solution:
(a) Since there are two conversion periods per year, then we have $m = 2$, and $i = 0.07/2 = 0.035$. Using Equation (4), we now
obtain:

$$\begin{aligned} r_E &= (1 + i)^m - 1 \\ &= (1 + .035)^2 - 1 \\ &= 0.071225 \\ &\approx 0.071 \quad \textbf{Solution} \end{aligned}$$

Thus, we see that the effective interest rate is about 7.1%. This means that if we were to invest an initial sum of money at the simple interest rate of 7.1%, then after one year, it would generate the *same* total amount as if the initial sum had been compounded semiannually at 7%.

EXERCISE SET #22

PART A

In Exercises 1–5, use the compound interest formula to calculate the future value A given that:

1. \$1200 is invested at 8% compounded quarterly for 2 years.
2. \$2500 is invested at 6% compounded semiannually for 3 years.
3. \$1800 is invested at 5% compounded monthly for 36 months.
4. \$8300 is invested at 10% compounded quarterly for 18 months.
5. \$6400 is invested at 12% compounded quarterly for 18 months.

In Exercises 6–10, use the present value formula to determine the initial principal which is required to generate each amount:

6. To generate \$1500 in 18 months at 4% compounded quarterly.
7. To generate \$2800 in 12 months at 3% compounded monthly.
8. To generate \$1280 in 6 months at 2% compounded quarterly.
9. To generate \$5000 in 24 months at 4% compounded quarterly.
10. To generate \$800 in 12 months at 2% compounded monthly.

PART B

11. A restaurant owner borrowed \$12000 at 8% compounded semiannually for 2 years. At the end of two years, how much money does the restaurant owner owe?
12. A financial adviser instructs his client to invest a sum of \$8200 at 6% compounded quarterly for 3 years. How much money will his client have at the end of this time period?
13. A certain bank has an interest rate of 8% compounded quarterly. What is the effective interest rate?
14. A credit union offers an interest rate of 6% compounded semiannually. Find the effective interest rate.
15. The Happy Savings Bank offers an interest rate of 6% compounded quarterly while the Peoples Credit Union offers an interest rate of 6.2% compounded semiannually. Determine the following:

    (a) the effective interest rate for the Happy Savings Bank.
    (b) the effective interest rate for the Peoples Credit Union.
    (c) which institution gives you the best deal.

16. An invester made a *net profit* of \$1800 by investing a sum of \$10,000 for a period of two years. If the interest was compounded quarterly, then what was the interest rate?
17. A certain credit card company imposes a finance charge of 1.2% per month on a customer's total unpaid balance. If Bob purchases a \$1200 stereo with his credit card, and doesn't make any payments for 9 months, then how much money will he owe at that time?
18. Using the same information as in Exercise-17, determine the effective interest rate which is charged by the credit card company on an unpaid balance.
19. A parent is planning to invest some money in preparation for his daughter's college education. If he wants to double his initial investment in 4 years, and his money is to be compounded annually, then what must be the *annual rate* of interest?
20. When her daughter was born, Alice opened a savings account for her daughter with an initial deposit of \$10,000. The bank payed an annual interest rate of 4.5% compounded semiannually. If Alice closes the account on her daughter's 18th birthday, then how much money will she have at that time?

23 ANNUITIES

In this lesson we'll examine a special type of financial transaction known as an annuity. An **annuity** is just a series of *equal* payments that are made over equally spaced intervals of time. Installment payments, mortgage payments, periodic bank deposits, etc., are all examples of annuities.

Each payment is called the **periodic rent** or **periodic payment,** and when each such payment is made at the *end* of each payment period, we shall refer to such an annuity as an **ordinary annuity.** The total amount of time over which these payments are made is called the **term** of the annuity.

We call the sum of all of the periodic payments as well as the interest which they accrue the **future value** of an annuity. As you will see, this is a particularly useful quantity, and a formula for its computation is given in the first theorem.

Theorem-1: (Future Value of an Ordinary Annuity)

If an ordinary annuity consists of a total of n equal payments of periodic rent R where the interest rate per conversion period is i, then its **future value** A is given by:

FUTURE VALUE OF AN ORDINARY ANNUITY

$$A = R\left[\frac{(1+i)^n - 1}{i}\right] \quad (1)$$

Example-1: Bob deposits $1000 per year for a total of four years at 8% compounded annually. Determine the following:

(a) the future value of the annuity
(b) the total interest earned

Solution:

(a) Here we have $i = 0.08$, $n = 4$, and $R = \$1000$. Using (1), the future value A is given by:

$$A = R\left[\frac{(1+i)^n - 1}{i}\right] = (\$1000)\left[\frac{(1+.08)^4 - 1}{.08}\right]$$
$$= (\$1000)(4.5061125)$$
$$= \$4506.11 \quad \textbf{Solution}$$

Thus, the future value of the annuity is \$4506.11, rounded to the nearest cent.

(b) The total interest earned is just the difference between the future value of the annuity and the sum of the periodic rents. That is:

$$\text{Interest} = A - (\text{sum of all periodic rents})$$

$$= \$4506.11 - \$4000.00$$
$$= \$506.11 \quad \textbf{Solution}$$

Another quantity which is of equal importance is the **present value** of an annuity. The present value of an annuity is defined to be the sum of all of the *present values* of the periodic rent payments made during the term of the annuity. The next theorem gives us a convenient way to calculate this quantity.

Theorem-2: (Present Value of an Ordinary Annuity)

The present value P of an ordinary annuity which consists of n equal periodic payments R at an interest rate i per conversion period is given by:

PRESENT VALUE OF AN ORDINARY ANNUITY

$$P = R\left[\frac{1-(1+i)^{-n}}{i}\right] \qquad (2)$$

Example-2: Determine the amount of money which should be invested *now* so that a retired person can receive annual payments of \$5000 over a period of four years. Assume that the investment will be compounded annually at 8%.

Solution:

(a) What we're really trying to find here is the present value of the annuity. Here, the periodic rent $R = \$5000$, the interest per conversion period $i = 0.08$, and $n = 4$. From (2), we have:

$$
\begin{aligned}
P &= (\$5000)\left[\frac{1-(1+.08)^{-4}}{.08}\right] \\
&= (\$5000)(3.3121266) \\
&= \$16,560.63 \quad \textbf{Solution}
\end{aligned}
$$

Thus, we see that \$16,560.63 invested *now* at an annual interest rate of 8% will allow for annual payments of \$5000 for the next four years.

Equation (2) can also help us in calculating loan payments and mortgages. Most loans are **amortized** or repaid according to an annuity. That is, the original loan amount as well as the interest are paid over n payment periods with a fixed rate of interest.

If we solve Equation (2) for R then we obtain an **amortization formula** which tells us what our periodic payments R will be if we are given the total amount of the loan. In doing so, we obtain:

AMORTIZATION FORMULA

$$
R = P\left[\frac{i}{1-(1+i)^{-n}}\right] \tag{3}
$$

where as before, it is assumed that the payments are made at the end of each payment period, and where:

R = periodic loan payment (periodic rent)
i = interest rate per conversion period
P = total amount of loan (present value)
n = total number of periodic payments

Let's see how this formula can be used; take a look at the last example of this lesson.

Example-3: Bob purchased a new Corvette at a cost of \$46,000 and he made a down payment of \$10,000. He took out a loan for the balance of \$36,000 at an annual interest rate of 8% for 5 years.

(a) What are Bob's monthly payments?
(b) What is the total amount of interest that he will pay after 5 years?

Solution:

(a) Here, we have:

$$i = \frac{.08}{12} = 0.00667$$
$$P = \$36000$$
$$n = 12 \cdot 5 = 60$$

Consequently, we get:

$$R = (\$36000)\left[\frac{(0.00667)}{1-(1.00667)^{-60}}\right]$$
$$(\$36000)(0.020278299)$$
$$= \$730.02 \quad \textbf{Solution}$$

Thus, Bob would have to make monthly payments of \$730.02 for five years in order to pay off the loan.

(b) In order to calculate the total interest that Bob must pay, observe that:

$$\text{Total Interest} = \text{Total Payments} - \text{Amount Borrowed}$$
$$= (\$730.02)(60) - (35000)$$
$$= \$8{,}801.20 \quad \textbf{Solution}$$

Thus, we see that Bob would pay a total amount of \$8,801.20 in interest on the loan.

EXERCISE SET #23

PART A

1. Alice deposits $500 per year for a total of four years at 8% compounded annually. Determine the following:

   (a) the future value of the annuity
   (b) the total interest earned

2. Bob deposits $1500 per year for a total of five years at 6% compounded annually. Determine:

   (a) the future value of the annuity
   (b) the total interest earned

3. A student devotes $50 per month to an investment fund that pays 4% compounded monthly. How much money will the student have after one year?
4. In preparing for his retirement, Bob decides to invest $1000 per year in a retirement fund which pays 8% interest annually. How much money will Bob have accumulated after 20 years?
5. Mr. Jones opened an IRA (Individual Retirement Account) at the age of 22 and decided to set aside $1000 per year over the next 40 years. If the IRA provides an interest rate is 8% compounded annually, then how much money will he have at age 62?
6. A business owner decides to invest a total of $25,000 at an interest rate of 12% compounded monthly. What equal monthly payments (periodic rent) will he be eligible to receive over the 10 year period?
7. A student invests $2500 at an interest rate of 8% compounded monthly. What equal monthly payments can the student receive over a 5 year period?
8. Determine the amount of money which should be invested *now* so that a person can receive annual payments of $4000 over a period of five years. Assume that the investment will be compounded annually at 8%.
9. A person planning to retire would like to receive annual payments of $5000 over a period of 10 years. Assuming an interest rate of 8% compounded annually, how much money should be invested now to ensure these payments?
10. The Smith family purchased a new home at a cost of $156,000, and they made a down payment of $40,000. They decided to take out a mortage loan for the balance at an annual interest rate of 8% for 30 years.

(a) What are their monthly mortgage payments?
(b) What is the total amount of interest that they will pay?

11. A student purchased a new sound system at a cost of \$1200, and she made a down payment of \$100. She took out a loan for the balance at an annual interest rate of 8% for one year. What are her monthly payments?
12. Ralph purchased a used car at a cost of \$8200, and he didn't make any down payment. He took out a loan at an annual interest rate of 7.5% for four years. What are Ralph's monthly payments?

PART B

13. Derive Equation (2a).
14. Derive Equation (3).
15. A restaurant owner purchased some equipment for his new kitchen at a cost of \$27,000. He made a down payment of \$3000 and took out a loan for the balance at 8% compounded monthly over a period of 5 years. What will his monthly payments be?

APPENDIX A

APPLIED PROBLEMS

The problems in this appendix have specific application to business related fields.

Percents, Decimals, and Fractions

In Problems 1-8 write each decimal as a percent.

1. 0.45
2. 0.85
3. 1.12
4. 1.25
5. 0.06
6. 0.07
7. 0.0025
8. 0.0015

In Problems 9-16 write each percent as a decimal.

9. 42%
10. 7.25%
11. 0.2%
12. 300%
13. 0.001%
14. 4.3%
15. 73.4%
16. 92%

In Problems 17-30 calculate the indicated quantity.

17. 15% of 1000
18. 20% of 500
19. 18.% Of 100
20. 10% of 50
21. 210% of 50
22. 135% of 1000
23. What percent of 80 is 4?
24. What percent of 60 is 5?
25. What percent of 5 is 8?
26. What percent of 25 is 45?
27. 8% of what number is 20?
28. 12% of what number is 25?
29. 15% of what number is 50?
30. 18% of what number is 40?

Sets, Unions, and Intersections

In Problems 32-34 use...

$A = \{x \mid x$ is a customer of IBM$\}$

$A = \{x \mid x$ is a secretary employed by IBM$\}$

$A = \{x \mid x$ is a computer operator at IBM$\}$

$A = \{x \mid x$ is a stockholder of IBM$\}$

$A = \{x \mid x$ is a member of the Board of Directors IBM$\}$

to describe each set in words

31. $A \cap E$
32. $B \cap D$
33. $A \cap D$
34. $C \cap E$

35. Motors Inc., Manufactured 325 cars with automatic transmissions, 216 with power steering, and 89 with both these options. How many cars were manufactured if every car has at least one option?

36. **Survey Analysis** In a survey of 75 college students, it was found that of the three weekly news magazines *Time, Newsweek* and *U.S. News and World Report:*

23	read *Time*
18	read *Newsweek*
14	read *U.S. News and World Report*
10	read *Time* and *Newsweek*
9	read *Time* and *U.S. News and World Report*
8	read *Newsweek* and *U.S. News and World Report*
5	read All three

(a) How many read none of these three magazines?
(b) How many read *Time* alone?
(c) How many read *Newsweek* alone?
(d) How many read *U.S. News and World Report* alone?
(e) How many read neither *Time* alone?
(f) How many read *Time* or *Newsweek* or both?

37. **Car Sales** Of the cars sold during the month of July, 90 had air conditioning, 100 had automatic transmissions, and 75 had power steering. Five cars had all three of these extras. Twenty cars had none of these extras. Twenty cars had only air conditioning; 60 cars had only automatic transmissions; and 30 cars had only power steering. Ten cars had both automatic transmission and power steering.

(a) How many cars had both power steering and air conditioning?
(b) How many had both automatic transmission and air conditioning?
(c) How many cars had neither power steering nor air conditioning?
(d) How many cars were sold in July?
(e) How many had automatic transmission or air conditioning or both?

38. **Survey Analysis** Of 100 personal computers users surveyed: 27 use IBM; 35 use Apple; 35 use AT&T; 10 use both IBM and Apple; 10 use both IBM and AT&T; 10 use both Apple and AT&T; 3 use all three; and 30 use another computer brand. How many people exclusively use one of the three brans mentioned, that is, only IBM or only Apple or only AT&T?

Functions

39. **Revenue** A parking lot charges by the hour as follows:

$$f(x) = 2 + 1.5(x-1) \quad \text{for } x \quad 1$$

Find the charge for 1 hour, 2 hours, 10 hours, and 24 hours. Try to state the parking rate in words.

40. **Revenue Function** A travel agency books a flight to Europe for a group of college students. The fare in a 200-passenger airplane will be $400 per student plus $2.00 per student for each vacant seat.

(a) Write the total revenue $R(x)$ as a function of empty seats, x.
(b) What is the domain of this function?
(c) Calculate $R(x)$ for 5, 10, 20, 40, and 100 empty seats.

41. **Cost and Revenue.** A revenue function is given a R=100x. The cost function is defined by C= 1000 + 80x. Draw R and C on the same coordinate system and determine for what production there is or is not a profit. Verify using graphing calculator.
(**Hint:** Profit = Revenue - Cost)

42. **Depreciation.** Let x in the equation y = -60x = 10,000 represent months and y represent the dollar value of a machine less depreciation.
(a) Prepare a table of values for x - 0, 5, 10, 15, 20, 25, 30, 40.
(b) Sketch the graph of the equation, letting x assume all real values greater than or equal to 0

43. **Supply and Demand.** Suppose the demand D (price per unit) for a certain item varies with the number of units(x) so that

$$Pd = 50 - \frac{5x}{2}$$

(a) What is the price when $x = 0$?
(b) What is the price when $x = 4$?
(c) What happens to the price when $x = 10$?
(d) Graph the equation.

44. Every Monday, a newsstand sells x copies of a weekly sports magazine for $2.50. The owner of the newsstand buys the magazines for $1.70 a copy, plus a delivery fee of $50.
(a) Write an equation that relates the profit, in dollars, to the number of copies sold; graph this equation.
(B) How many copies must be sold to make a profit?
(c) What will the profit be if 200 copies are sold?

45. **Demand Function.** Suppose that the demand function (in price per unit) for a certain item is given by

$$p = D(x) = 50 - \frac{5x}{2}$$

when x units are demanded by the consumer at price p=D(x).

(a) What is $D(x)$ when $x = 0$?
(b) What is $D(x)$ when $x = 4$?
(c) What happens to $D(x)$ when x = 10?
(d) Sketch the graph of $D(x)$.

46. **Supply Function.** Suppose that the supply for the item in Exercise 33 is given by

$$p = S(x) = \frac{5x}{6}$$

where $p = S(x)$ is the price per unit of an item a which the seller is willing to supply x units.

(a) What is $S(x)$ when $x = 3$? When $x = 9$?
(b) Sketch the graph of $S(x)$ on the same axis system that you used in Exercise 33.
(c) Estimate the point of equilibrium from the intersection of the graphs.
(d) Estimate the equilibrium price.
(e) When is price greater that demand?

47. **Demand and Supply Function.** The supply function $S(x)$ and the demand function $D(x)$ for a certain commodity in terms of units available, x, at price p, are

$$p = S(x) = 400 - \frac{5x}{2} \text{ and } p = D(x) = \frac{5x}{2}$$

(a) Graph $S(x)$ and $D(x)$ on the same axis.
(b) Find the equilibrium point.
(c) Find the equilibrium price.
(d) When is the supply less than demand?

48. **Commission.** A man is trying to decide between two positions. The first pays \$225 per week plus 5% commission on gross sales. The second pays only 9% on gross sales. Graph the two pay functions and find where they are equivalent.

49. **Break-Even Point.** A producer knows that she can sell as many items at \$0.25 each as she can produce in a day. If her cost is $C = \$0.20x + \$70$, find her break-even point.

50. **Break-Even Point.** A firm knows that it can sell as many items \$01.25 each as it can produce in a day. If the cost is $C = \$0.90x + \$150$, find the break-even point.

51. **Break-Even Point.** A firm knows that it can sell as many items \$01.25 each as it can produce in a day. If the cost is $C = \$0.90x + \$150$, find the break-even point.

52. **Cost of Renting a Truck.** The cost of renting a truck is \$280 per week plus a charge of \$0.20 per mile driven. Write an equation that relates the cost C for a weekly rental in which the truck is driven x miles.

53. **Wages of a Car Salesman.** Dan receives \$300 per week for selling cars. As part of his weekly salary he also receives 10% of the sales he generates. Write an equation that relates Dan's weekly salary S when he generates sales of x

54. **Break-Even Point.** A manufacturer produces items at a daily cost of \$0.75 per item and sells them for \$1 per item. The daily operational overhead is \$300. What is the break-even point? Graph your result.

55. **Break-Even Point.** If the manufacturer of Problem 11 is able to reduce the cost per item to \$0.65, but with a resultant increase to \$350 in operational overhead, is it advantageous to do so? Graph your result.

56. **Profit from Selling Newspapers.** Each Sunday, a newspaper agency sells x copies of a certain newspaper for \$1.00 per copy. The coat to the agency for each newspaper is \$0.50. The agency pays a fixed cost for storage, delivery, and so on, of \$100 per Sunday. How many newspapers need to be sold for the agency to break-even?

57. **Profit from Selling Newspapers.** Repeat Problem 13 if the cost to the agency is \$0.45 per copy and the fixed cost is \$125 per Sunday.

58. **Investment Problem.** Mr. Nicholson finds after 2 years that because of inflation he now needs \$12,000 per year in supplementary income. How should he transfer his funds to achieve this amount? (Use the data from Problem 17.)

59. **Marketing Survey.** A recent survey found that 60% of the people in a given community drink a certain cola and 40% drink other soft drinks; 15% of the people interviewed indicated that they drink both cola and other sift drinks. What percent of the people drink the cola or other soft drinks?

60. **Forecasting.** In a survey of presidents of leading banks by an economics consulting group, the following information was obtained relative to their forecast for next year:

    65% expect higher inflation
    15% expect a recession
    5% expect both higher inflation and a recession
    75% expect higher interest rates
    50% expect both higher inflation and higher interest rates
    10% expect higher interest rates and a recession
    3% expect higher inflation, higher interest rates, and a recession

    What is the probability that a bank president selected at random would forecast

    (a) no recession or lower interest rates?
    (b) no increase in inflation and no increase in interest rates?
    (c) no recession, or no increase in interest rates, or no increase in inflation?

61. **Reading Habits.** A survey of 100 people in a library revealed the following:

    40 read the *Wall Street Journal*
    30 read *National Geographic*
    25 read *Sports Illustrated*
    15 read the *Wall Street Journal* and *National Geographic*
    12 read the *Wall Street Journal* and *Sports Illustrated*
    10 read *National Geographic* and *Sports Illustrated*
    4 read all three

    If a person surveyed is selected at random, what is the probability that the person

    (a) reads only two magazines?
    (b) reads only one magazine?
    (c) reads *Sports Illustrated* and *National Geographic* but not the *Wall Street Journal*?
    (d) does not read any of the magazines?

62. **Consumer Survey.** A survey was taken of 120 people concerning their preferences for toothpaste samples they were asked to try. The samples were labeled A, B, C. The results were as follows:

    22 liked A
    25 liked B
    26 liked C

    15 liked A and B
    10 liked A and C
    12 liked B and C
    8 liked all three

    If one of those surveyed chosen at random, find the probability that the person
    (a) liked at least one of the brands(b) liked A and B but not C
    (c) did not like any brand
    (d) liked only one brand

63. **Advertising.** In Atlanta, 600,000 people read newspaper A, 450,000 read newspaper B, and 160,000 read both newspapers. How many read either newspaper A or newspaper B?

64. **Quality Control.** A machine is assembled using components A and B. The two components are each built in separate fabricating plants. Experience indicates that the probability that A is defective is .01 and the probability that B is defective is .05. (**Hint:** Since A and B are fabricated in different plants, whether A is good or defective is independent of the quality of B.)

    (a) What is the probability that both components are defective?
    (b) What is the probability that both components are good?

65. **Quality Control.** You know that 4% of all light bulbs produced by a given company weigh less than specifications, and 2% of all bulbs are both defective and less than specifications. What is the probability that a light bulb selected at random is defective, if you know that it weighs less than specifications?

66. **Quality Control.** A manufacturer receives a shipment of 20 articles. Unknown to him, 6 are defective. He selects 2 articles at random and inspects them. What is the probability that the first is defective and the second is satisfactory?

67. Plants A and B manufacture tires for a warehouse. The warehouse receives 40% of its tires from plate A and 60% from plant B. Of the tires made by plant A,
90% are good and 10% have some defect. Of the tires made by plant B, 85% are good and 15% have some defect. Suppose that one of the plants is chosen at random and two tires from that plant are selected.

    (a) What is the probability that neither tires are defective?
    (b) What is the probability that both tires are good if it is known that they came from plant B?

68. **Product Testing.** The probability that a new, long-lasting AA battery will last at least 15 hours is .80 and the probability that it will last at least 20 hours is .15. A battery is selected at random and after 15 hours is still good. What is the probability that it will last at least 20 hours?

69. **Contracts.** A corporation prepares a bid on a job at a cost of $7000. They estimate that if they get the job, they will make $250,000 in profits. If the probability of getting the job is .4 what is their expected profit or loss?

70. **Contracts.** Find the expected values for the following contracts.

    (a) Estimated profit $5000, cost of proposal $500, probability of winning 1/8
    (b) Estimated profit $1000, cost of proposal $200, probability of winning 1/5
    (c) Estimated profit 10,000, cost of proposal $600, probability of winning 1/3

71. **Sales.** During a sale, an appliance dealer offers a chance on a $1200 motorcycle for each refrigerator sold. If he sells his refrigerators at $25 more than other dealers, and if he sells 120 units during the sale, what is the expected value of a purchase to the consumer?

72. Compute the simple interest when the principal, rate and time of the loan are given.

    (a) $P = \$500,\ r = 0.08,\ t = 2$ years
    (b) $P = \$300,\ r = 0.03,\ t = 4$ years
    (a) $P = \$500,\ r = 0.04,\ t = 5$ years

73. What is the amount to be repaid in (a), (b), and (c) of Exercise 1?

74. Find the interest and the amount of a loan for \$3000 borrowed for 2 years at 6% simple interest.

Find the compound interest and compound amount for the investments in Exercises 75 through 77

75.
    (a) \$5000 at 8% compounded annually for 3 years
    (b) \$5000 at 8% compounded semiannually for 3 years
    (c) \$5000 at 8% compounded quarterly for 3 years

76.
    (a) \$2000 at 8% compounded annually for 4 years
    (b) \$2000 at 8% compounded semiannually for 4 years
    (c) \$2000 at 8% compounded quarterly for 4 years

77.
    (a) \$3000 at 6% compounded annually for 6 years
    (b) \$3000 at 6% compounded semiannually for 6 years
    (c) \$3000 at 6% compounded quarterly for 6 years

78. Compute the amount to be repaid when the principal, simple interest rate and time of the loan are given.

    (a) $P = \$4000,\ r = 0.06,\ t = 2$ years
    (b) $P = \$3000,\ r = 0.06,\ t = 6$ months
    (a) $P = \$100,\ r = 0.08,\ t = 3$ years

79. Find the simple-interest rate on the loan when the principal, the amount repaid, and the term of the loan are given.

    (a) Principal = \$3500, amount repaid in 2 years, \$4130
    (b) Principal = \$1000, amount repaid in 120 days, \$1046.67
    (c) Principal = \$500, amount repaid in 45 days, \$510

Find the present value of the money In Exercises 80 through 83

80. \$5000 due in 5 years if money is worth 6% compounded annually
81. \$6000 due in 5 years 6 months if money is worth 8% compounded semiannually
82. \$7000 due in 4 years if money is worth 6% compounded semiannually
83. \$8000 due in 6.5 years if money is worth 7% compounded annually
84. How long will it take for \$125 to amount to \$375 at 6% interest compounded annually?
85. How many years will it take to double \$1000 at 4% interest compounded semiannually?
86. **Interest.** Find the interest on \$2000 borrowed for 8 months at 10% simple interest.

87. **Interest.** Find the amount of a 90-day, $1500 loan at 6% simple interest. (Use a banker's year of 360 days in the computation.)
88. **Savings.** How much should parents invest for their daughter at 6% interest compounded semiannually in order to have $5000 at the end of 20 years?.
89. 8% compounded semiannually
90. 6% compounded semiannually
91. 8% compounded monthly
92. 10% compounded monthly
93. Find the accumulated value of $10,000 invested for 10 years at 4% interest compounded daily.
94. How many years will it take money to double if 8% interest is compounded daily?
95. **Investments.** A man is to receive $1000 at the end of each year for 5 years. If he invests each year's payment at 8% compounded annually, how much will he have at the end of 5 years?
96. **Investments.** Suppose that you deposit $500 each 6 months in a credit union that pays 8% interest compounded semiannually. How much would you have after 5 years?

97. **Payments.** Compute the monthly payment necessary to finance a used car for $3500 at 6% interest compounded monthly for 3 years.

98. **Payments.** Find the payment necessary each quarter for 2 years to amortize a debt of $2000 at 8% interest compounded quarterly.

For problems 99-101 zero coupons bonds *are used. A zero coupon bond is a bond that is sold now at a discount and will pay its face value at some time in the future when it matures; no interest payments are made.*

99. Tami's grandparents are considering buying a $40,000 face value zero coupon bond at birth so that she will have enough money for her college education 17 years later. If money is worth 8% compounded annually, what should they pay for the bond?

100. How much should a $10,000 face value zero coupon bond, maturing in 10 years, be sold for now if its rate of return is to be 8% compounded annually?

101. If you pay $12,485.52 for a $25,000 face value coupon bond that matures in 8 years, what is your annual compound rate of return?

102. A bank advertises that it pays interest on saving accounts at the rate of 6.5% compounded daily. Find the effective rate if the bank uses (a) 360 days or, (b) 365 days in determining the daily rate.

Problems 103 and 104 require logarithms

103. How many years will it take for an initial investment of $10,000 to grow to $25,000? Assume a rate of interest of 6% compounded daily.

104. How many years will it take for an initial investment of $25,000 to grow to $80,000? Assume a rate of interest of 7% compounded daily

105. The grand prize in the Illinois lottery is $6,000.000 paid out in 20 equal yearly payments of $300,000 each. How much should the state deposit in an account paying 8% compounded annually to achieve this goal?

106. Mike and Yola have just purchased a town house for $76,000. They obtain financing with the following terms: a 20% down payment and the balance to be amortized over 30 years at 9%.

    (a) What is their down payment?
    (b) What is the loan amount?
    (c) How much is their monthly payment on the loan?
    (d) How much total interest do they pay over the life of the loan?
    (e) If they pay an additional $100 each month toward the loan, when will the loan be paid?
    (f) With the $100 additional monthly payment, how much total interest is paid over the life of the loan?

107. **House Mortgage.** Mr. And Mrs. Ostedt have just purchased an $80,000 home and made a 25% down payment. The balance can be amortized at 10% for 25 years.

    (a) What are the monthly payments?
    (b) How much interest will be paid?
    (c) What is their equity after 5 years?

108. **House Mortgage.** A mortgage of $125,000 is to be amortized at 9% per annum for 25 years. What are the monthly payments? What is the equity after 10 years?

APPENDIX B

USING THE TI-83 GRAPHING CALCULATOR

Basics

Turning the Calculator ON and OFF

The [ON] button is in the lower left corner of the calculator. To turn the calculator **OFF**, you need to press two buttons. First press the [2nd] key in the upper left corner, then press the [ON] button. Notice the word **OFF** above the [ON] button. To use any of the yellow functions printed above the keys, press the [2nd] key first, then the key just below the function you want. To use any of the green functions or letters printed above the keys, first press the [ALPHA] key.

Darkening and Lightening the Screen

You will need to adjust your screen brightness depending on lighting and the condition of your batteries. To darken the screen, first press the [2nd] key and then the [▲] key. You may need to repeat this key combination several times. To lighten the screen, press the [2nd] key and the [▼] key alternately and repeatedly, until the screen lightens. If you still have trouble adjusting the brightness, your batteries may need replacing. After you have replaced weak batteries, adjust the screen brightness as necessary.

Stopping the Machine

When you want to stop using a function or menu, you need to **QUIT**. Press [2nd], then the [MODE] key. This combination gives you the **QUIT** function.

Editing

Press the [CLEAR] key to clear one line of instructions. If the cursor is on a blank line on the home screen, press the [CLEAR] key to clear the entire screen. Using the arrow keys to first highlight an entry, press the [DEL] key to remove only that one entry. To insert within a previously written line, use the arrow keys and then the [2nd], **INS** combination. To recall your last command for editing or re-execution, use the [2nd], [ENTER] combination

Screen Display

If your screen looks "**w***e**ird*," check settings under [MODE] and also under [2nd] **FORMAT** The settings usually used are listed at the left side of the screen and should be highlighted. Use the arrow keys to highlight the desired settings, and the [ENTER] key to make changes. (See page 110 for more details.)

Using the FINANCE Features on the TI–83

First set the calculator to two decimal places of accuracy by pressing **MODE** and making the the following choice.

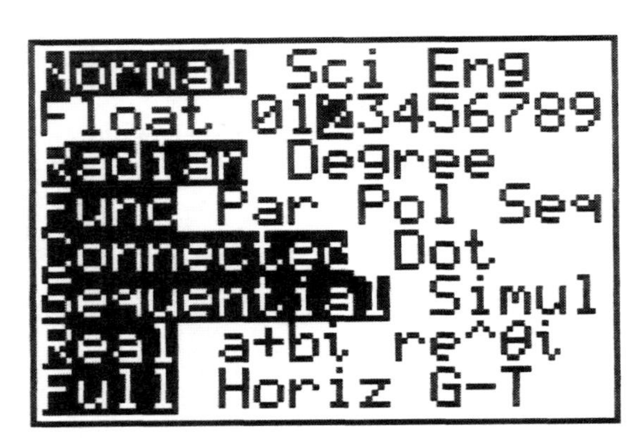

Now press **APPS** and select **1:Finance** (for the TI-83 Plus) or **2nd** and $\mathbf{x}^{-1}$ (this is the **FINANCE** key) (for the TI-83). Select **CALC** and **1:TVM Solver**.

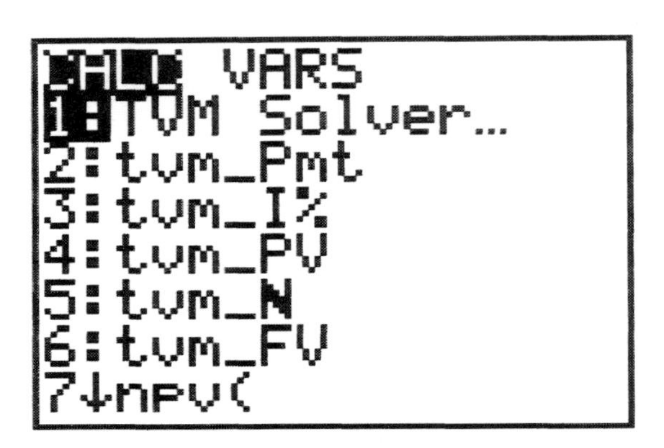

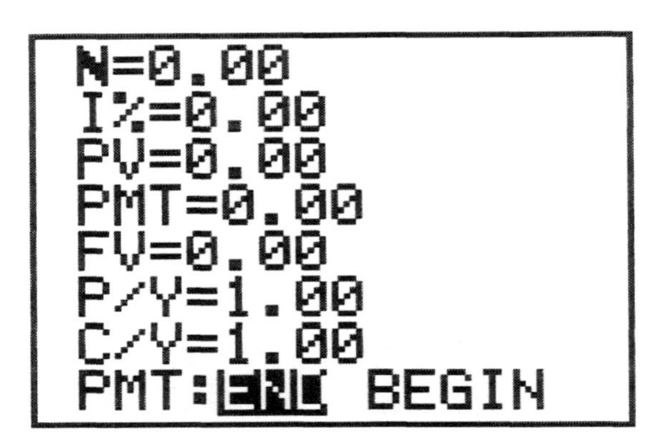

Note:

N = total # of payment periods

I% = annual interest rate

PV= present value

PMT= payment amount

FV = future value

P/Y= # of payment periods per year

C/Y= # of compounding periods per year

Ex. 1

What is the future value of $25,000 invested at 8.6% compounded quarterly for 6 years?

First we need to identify the values of the following variables:

N = 24	(6 years times 4 quarters per year)
I% = 8.6	(note this is entered as a %)
PV= -25000	(this is treated as a cash outflow)
PMT= 0	(there are no quarterly payments)
FV =	(this is our unknown)
P/Y= 4	(4 payment periods per year)
C/Y= 4	(4 compounding periods per year)

Now press **APPS** and select **1:Finance** (for the TI-83 Plus) or **2nd** and $\mathbf{x^{-1}}$ (this is the **FINANCE** key) (for the TI-83). Select **CALC** and **1:TVM Solver**. Enter the above information using the arrow keys.

```
N=0.00
I%=0.00
PV=0.00
PMT=0.00
FV=0.00
P/Y=1.00
C/Y=1.00
PMT:END BEGIN
```

```
N=24.00
I%=8.60
PV=-25000.00
PMT=0.00
FV=0.00
P/Y=4.00
C/Y=4.00
PMT:END BEGIN
```

Move the cursor to FV and solve for this value by pressing **ALPHA** (**GREEN** key) and **ENTER** (this is the **SOLVE** key).

```
N=24.00
I%=8.60
PV=-25000.00
PMT=0.00
FV=■.00
P/Y=4.00
C/Y=4.00
PMT:END BEGIN
```

```
 N=24.00
 I%=8.60
 PV=-25000.00
 PMT=0.00
▪FV=41654.40
 P/Y=4.00
 C/Y=4.00
 PMT:END BEGIN
```

Notice the small mark next to FV denoting that this is the quantity that we solved for. The other values were inputs.

FV = $41,654.40

Ex. 2

What is the amount of money required to generate a future value of $36,000 if the original amount is invested at 12% compounded semi-annually for 5 years?

First we need to identify the values of the following variables:

N = 10	(5 years times 2 periods per year)
I% = 12	(note this is entered as a %)
PV=	(this is our unknown)
PMT= 0	(there are no payments)
FV = 36000	(this is treated as a cash inflow)
P/Y= 2	(2 payment periods per year)
C/Y= 2	(2 compounding periods per year)

Now press **APPS** and select **1:Finance** (for the TI-83 Plus) or **2nd** and $\mathbf{x^{-1}}$ (this is the **FINANCE** key) (for the TI-83). Select **CALC** and **1:TVM Solver**. Enter the above information using the arrow keys.

```
N=10.00
I%=12.00
PV=0.00
PMT=0.00
FV=36000.00
P/Y=2.00
C/Y=2.00
PMT:END BEGIN
```

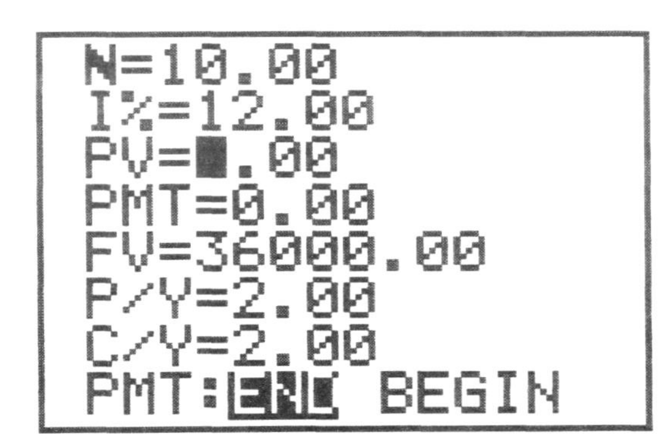

Move the cursor to PV and solve for this value by pressing **ALPHA** (**GREEN** key) and **ENTER** (this is the **SOLVE** key).

```
N=10.00
I%=12.00
▪PV=-20102.21
PMT=0.00
FV=36000.00
P/Y=2.00
C/Y=2.00
PMT:END BEGIN
```

Thus of final result is:

PV = $20,102.21

Ex. 3

At what annual interest rate, compounded quarterly, will an investment of $4,000 generate an amount of $10,000 in 10 years?

Using the TI-83, we press **MODE** and change to **Float** .

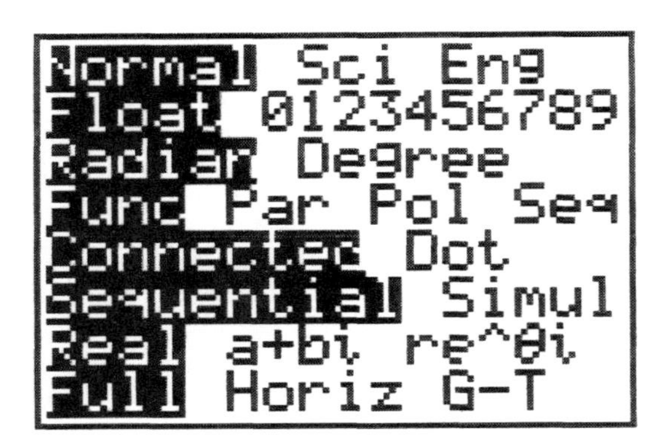

Using the TVM Solver we first note that

N = 40	(10 years times 4 periods per year)
I% =	(this is our unknown)
PV= -4000	(this is treated as a cash outflow)
PMT= 0	(there are no payments)
FV = 10000	(this is treated as a cash inflow)
P/Y= 4	
C/Y= 4	

Enter the inputs from above and move the cursor to I% and press **2nd** and **ENTER** (**SOLVE**).

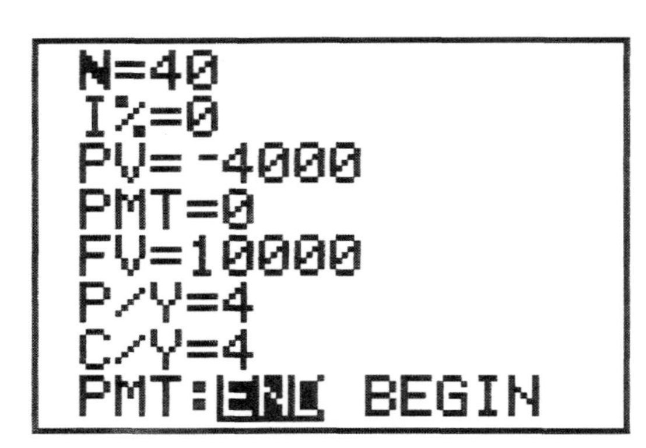

N=40
▪ I%=9.268661879
PV=-4000
PMT=0
FV=10000
P/Y=4
C/Y=4
PMT:END BEGIN

Thus our answer is

I%=9.268661879%

Ex. 4

How long will it take for a present value of $10,000 to become a future value of 100,000 at 18% compounded monthly?

Let's go directly to the TVM Solver. Again we determine the inputs and the unknown.

N = (this is our unknown in terms on months)
I% = 18
PV= -10000 (this is treated as a cash outflow)
PMT= 0 (there are no payments)
FV = 100000 (this is treated as a cash inflow)
P/Y= 12
C/Y= 12

```
N=0
I%=18
PV=-10000
PMT=0
FV=100000
P/Y=12
C/Y=12
PMT:END BEGIN
```

```
N=
I%=18
PV=-10000
PMT=0
FV=100000
P/Y=12
C/Y=12
PMT:END BEGIN
```

```
■N=154.6541086
I%=18
PV=-10000
PMT=0
FV=100000
P/Y=12
C/Y=12
PMT:END BEGIN
```

So our answer could be expressed as

N=154.65 months

To get the result in years we divide by 12 or we can use the TI-83 financial variables feature. First press 2nd and MODE (QUIT key) which gets us back to the home screen.

Now press **APPS** and select **1:Finance** (for the TI-83 Plus) or **2nd** and $\mathbf{x^{-1}}$ (for the TI-83) and select **VARS** and then select **1:N**.

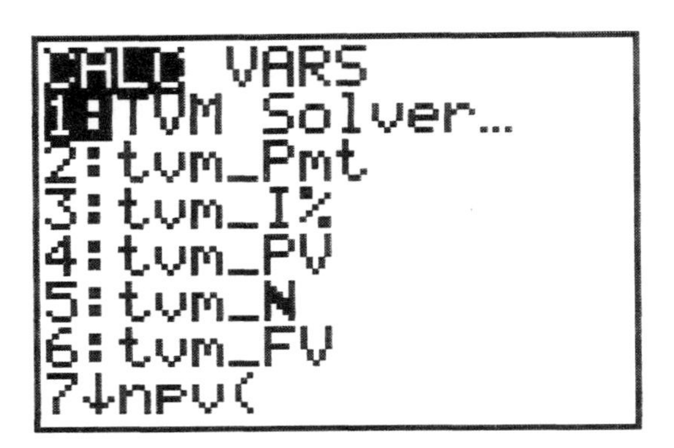

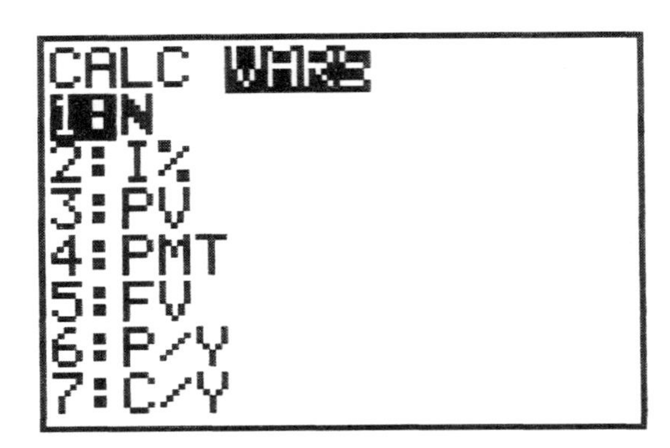

This pastes N onto the home screen. Divide this by 12 and you're done.

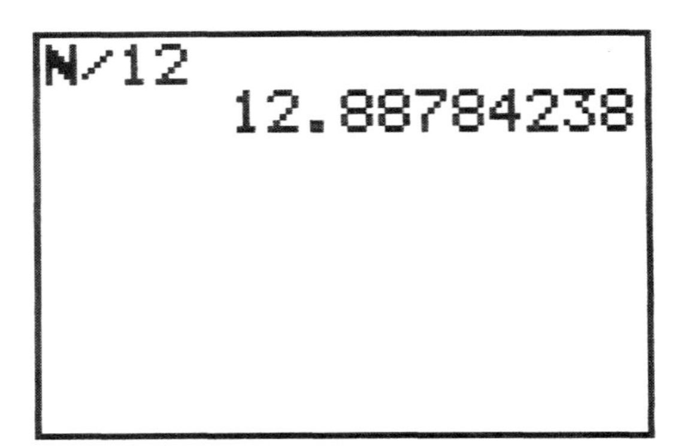

Thus our final result is

T= 12.89 years

Effective Rate of Interest (ERI)

The TI-83 has the capability of computing effective rates of interest given nominal rates and the number of compounding periods. Similarly we can find nominal rates of interest given effective rates of interest and the number of compounding periods. This will be done from the **FINANCE CALC** menu by selecting **B:▶Nom** and **C:▶Eff**.

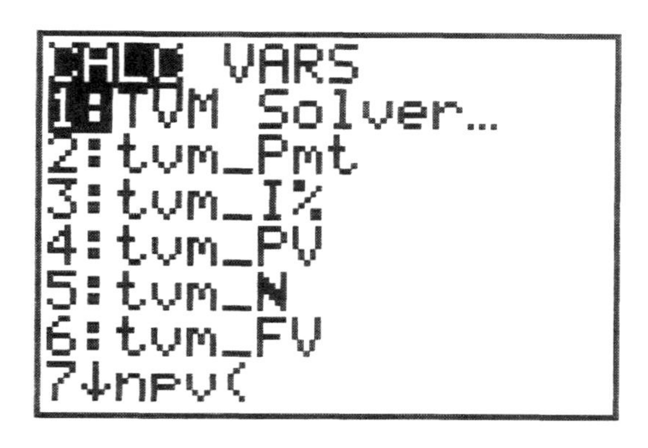

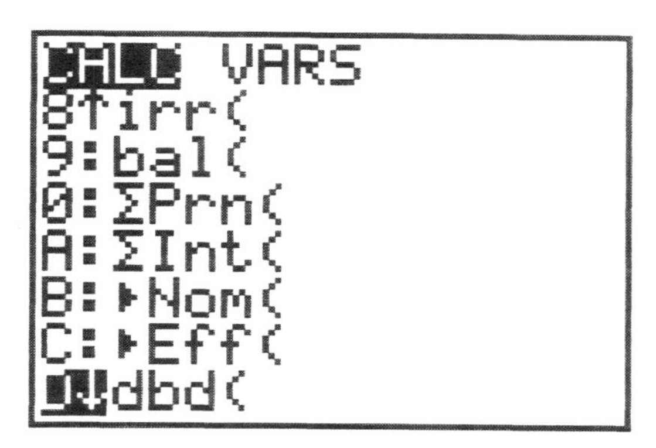

The general form for these features is as follows:

▶Eff (*nominal rate, number of compounding periods*)
▶Nom (*effective rate, number of compounding periods*)

Ex.5
Find the ERI for 8% compounded quarterly for 1 yr.

Starting at the Home Screen, press **APPS** and select **1:Finance** (for the TI-83 Plus) or **2nd** and **x^{-1}** (for the TI-83), and select **CALC** and **C:▶Eff.** This pastes the Eff feature into your home screen. Type in the nominal rate and the number of compounding periods. Then press **ENTER**.

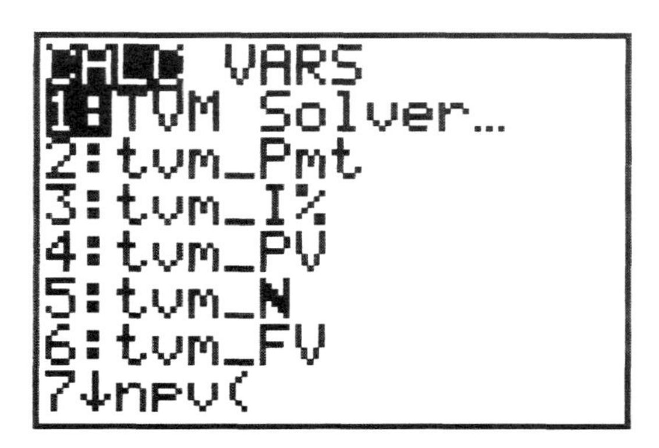

▸Eff(

Ex.6

Find the nominal rate for an effective rate of 6.25% compounded quarterly.

Starting at the Home Screen, press **APPS** and select **1:Finance** (for the TI-83 Plus) or **2nd** and **x**$^{-1}$ (for the TI-83), and select **CALC** and **B:▶Nom.** This pastes the Nom feature into your home screen. Type in the effective rate and the number of compounding periods. Then press **ENTER**.

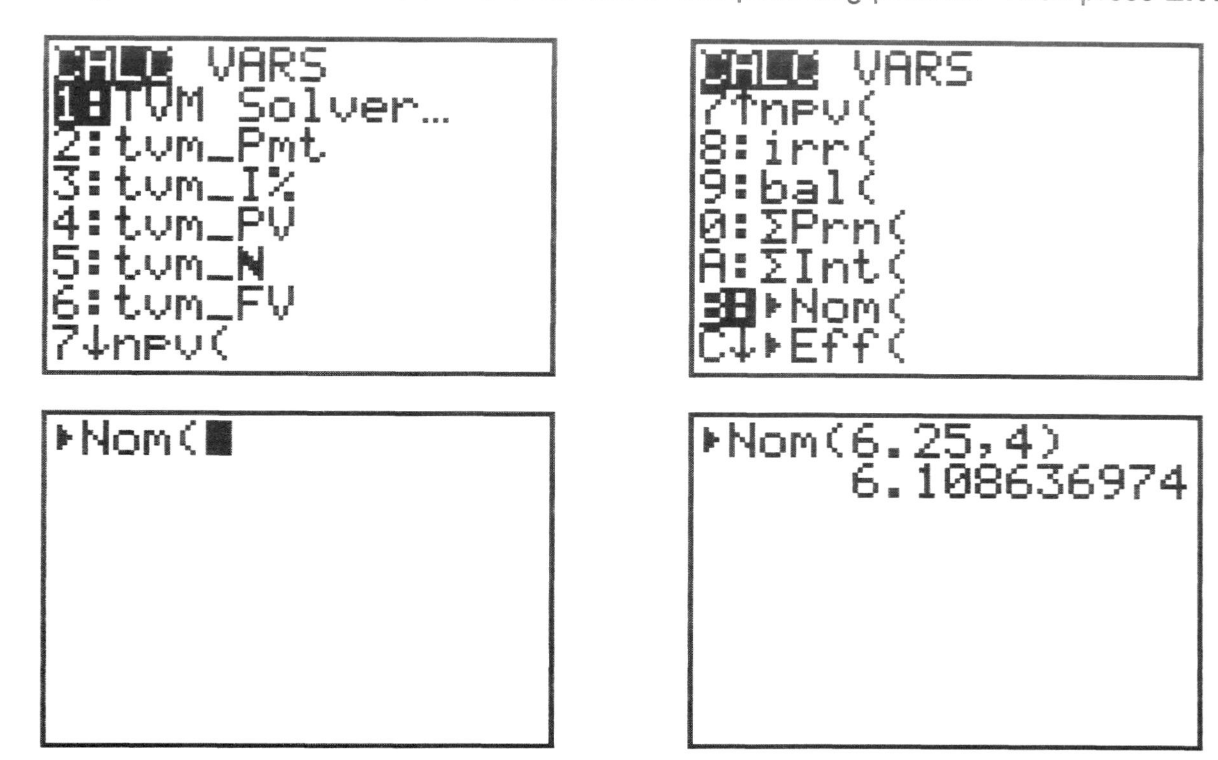

Mortgage Calculations on the TI-83

Recall that we access the TVM Solver by pressing **APPS** and select **1:Finance** (for the TI-83 Plus) or **2nd** and x^{-1} (for the TI-83). Select **CALC** and **1:TVM Solver**.

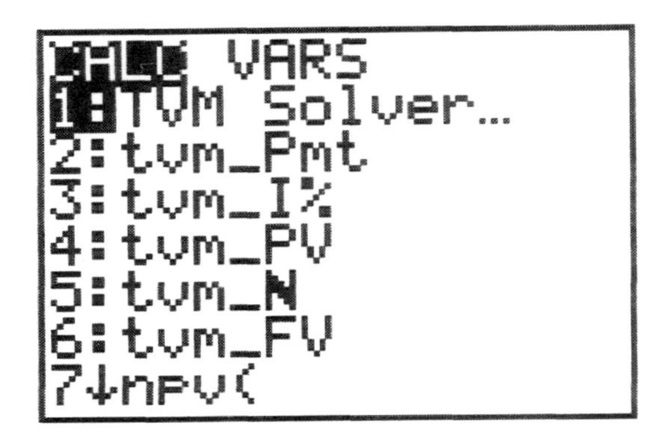

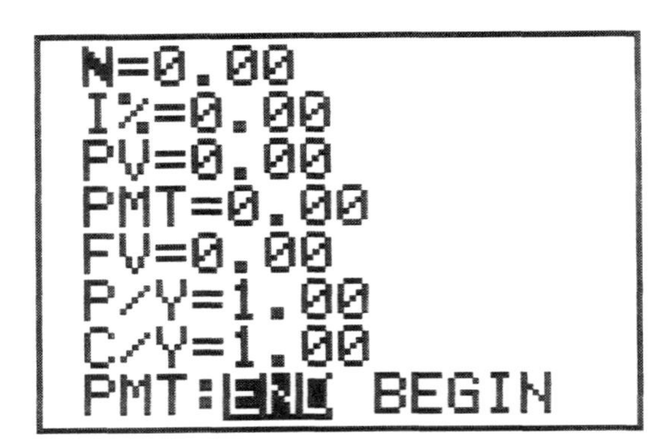

Note:

N = total # of payment periods

I% = annual interest rate

PV= present value

PMT= payment amount

FV = future value

P/Y= # of payment periods per year

C/Y= # of compounding periods per year

In addition note that **PMT:END** specifies an ordinary annuity, where payments occur at the end of each payment period. Most loans are in this category. **PMT:BEGIN** specifies an annuity due, where payments occur at the beginning of each payment period. Most leases are in this category.

<u>Ex.7</u>

A home is priced at $100,000. A downpayment of $25,000 is to be made, with the remainder financed at 10.75% for 30 years. **Find the monthly payment.**

N = 360 (30 years times 12 months)

I% = 10.75

PV= 75000

PMT= this is our unknown

FV = 0

P/Y= 12

C/Y= 12

After changing to 2 decimal places of accuracy (MODE), enter in the inputs as shown below.

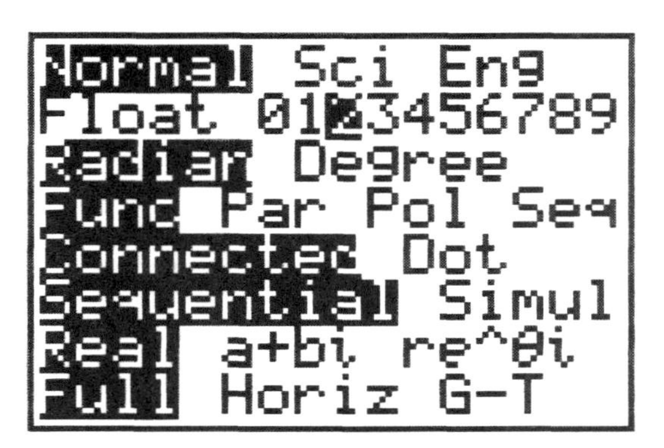

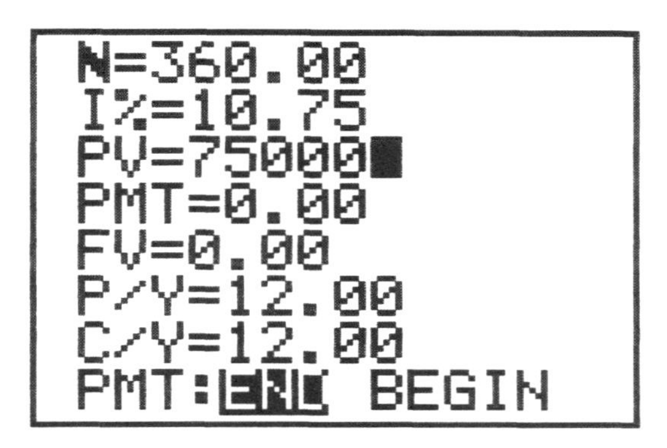

Move the cursor to PMT and press **ALPHA** (**GREEN** key) and **ENTER** (**SOLVE** key). The result is **$700.11** .

N=360.00
I%=10.75
PV=75000.00
PMT=0.00
FV=0.00
P/Y=12.00
C/Y=12.00
PMT:END BEGIN

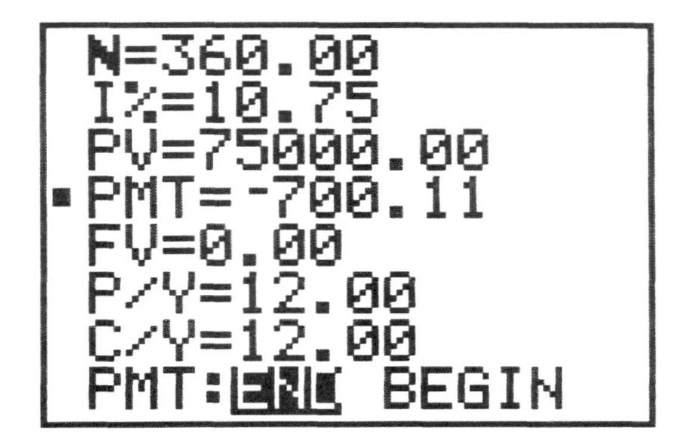

Ex. 7 continued

Find the Principal and Interest for the 1st, 10th, and 360th payment. Find the balance after the 1st, 10th, and 360th payment. Find the total principal and total interest paid over the 360 payments.

The three features we will need to access are found in the **FINANCE CALC** menu by scrolling down to 9:bal, 0:ΣPrn, A:ΣInt .

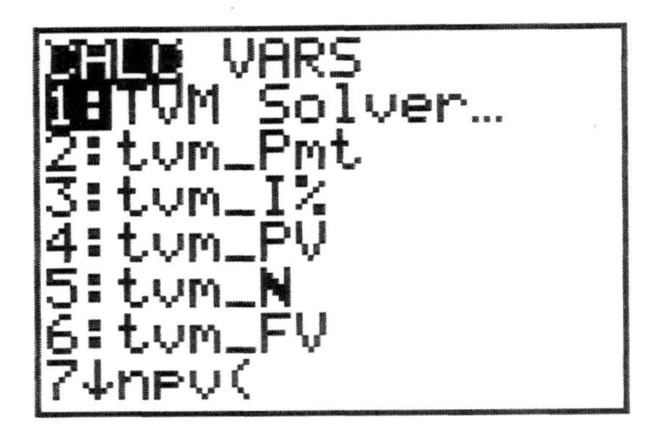

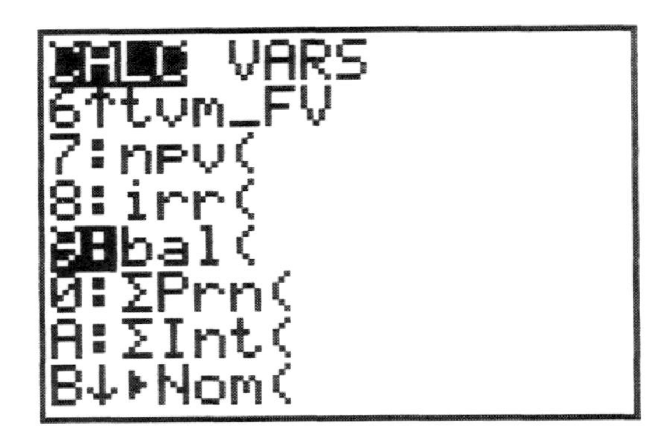

bal	bal(number of payment) computes the balance
ΣPrn	ΣPrn(payment x,payment y) computes the sum of the principal between payment x and payment y
ΣInt	ΣInt(payment x,payment y) computes the sum of the interest between payment x and payment y

The balances after the 1st, 10th, and 360th payments

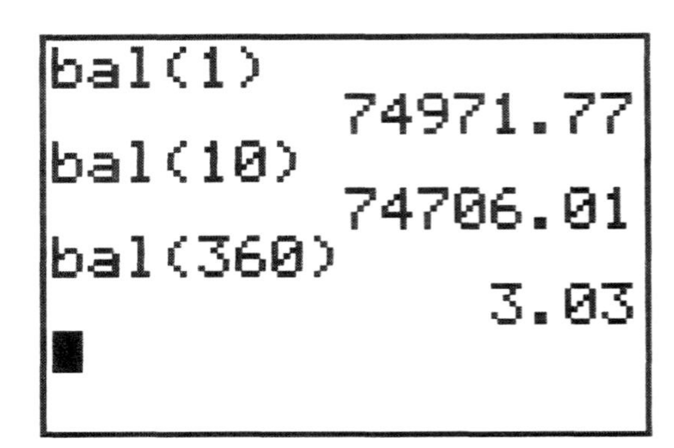

We find the principle and interest for the 1st, 10th, and 360th payment, as well as the total amount of principle and interest paid through 360 payments, as follows:

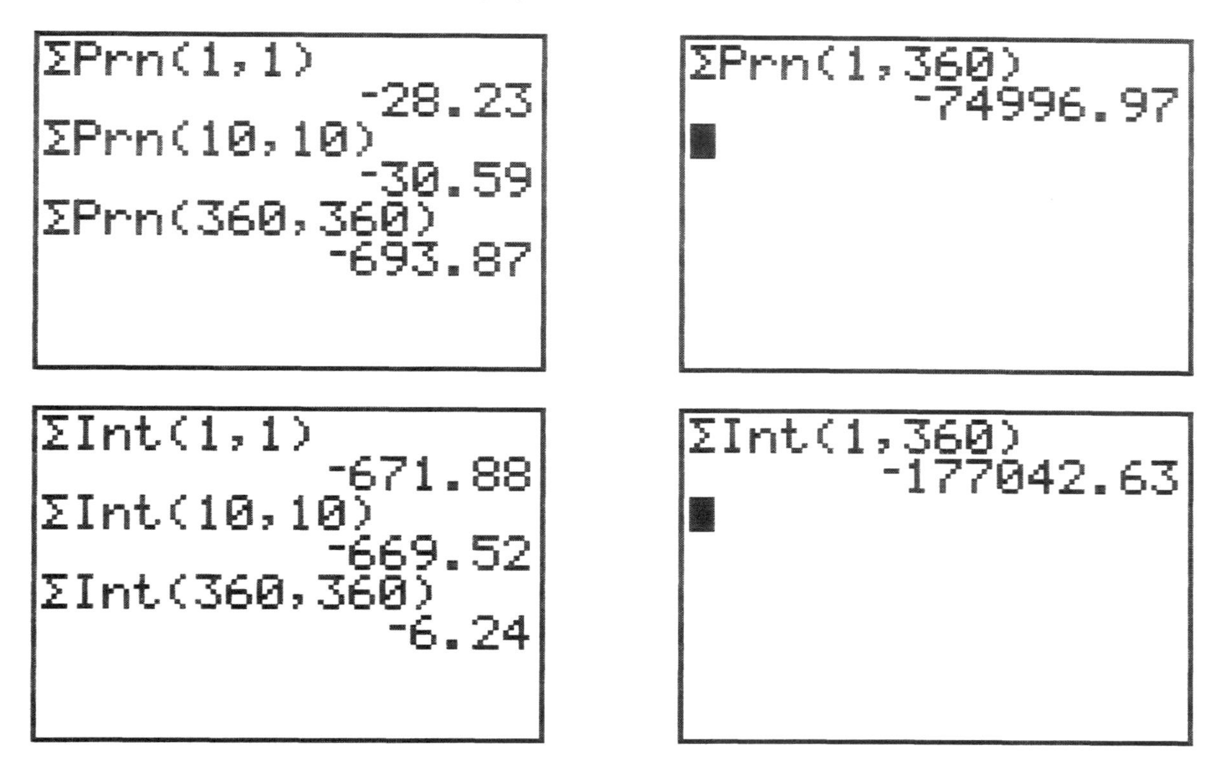

This tells us that the first payment will contain **$28.23** in **principal**, **$671.88** in **interest**, and a remaining **balance** of **$74,971.77.**

This tells us that the 10th payment will contain **$30.59** in **principal**, **$669.52** in **interest**, and a remaining **balance** of **$74,706.01 .**

This tells us that the 360th payment will contain **$693.87** in **principal**, **$6.24** in **interest**, and a remaining **balance** of **$3.03.**

ANSWERS TO SELECTED PROBLEMS

EXERCISE SET #1

1. {1,2,3,4}
3. {4,6,8}
5. {11,13,15,17,. . .}
7. {2,3,5,7,11,13,17,19,23,29}
9. False
11. True
13. True
15. $\{x \mid x \in N$ and x is odd, where $3 < x < 12\}$
17. $\{x \mid x \in N$ and x is an odd number where $x > 10\}$
19. $\{x \mid x \in N$ and x is an even number where $x < 9\}$
21. $\{x \mid x$ is a prime number and $x < 10\}$
23. $\{x \mid x \in N$ and $2 < x < 13\}$
25. $\{x \mid x \in N$ and $x = 5k$ for some natural number $k\}$

EXERCISE SET #2

1. True
3. True
5. False
7. True
9. Infinite
11. Finite
13. Finite
15. Equivalent
17. Both
19. $\#(A) = 4$
21. $\#(C) = 1$

EXERCISE SET #3

1. True
3. True
5. False
7. True
9. True
11. True
13. True
15. True
17. True
19. True
21. $P(A) = \{\emptyset, \{1, b, 2\}, \{1\}, \{b\}, \{2\}, \{1, b\}, \{1, 2\}, \{b, 2\}\}$
23. $2^7 = 128$ ways

EXERCISE SET #4

1. $\{a, b, c, d\}$
3. $\{e, f\}$
5. $\{a, b\}$
7. $\{b, c, d\}$
9. $\{a, d, e, f\}$
11. $\{b, c, d, e\}$
13. $\emptyset$
15. $\{e, f\}$
17. False
19. False
21. False
23. True

24b. $\{e\}$

EXERCISE SET #5

1.

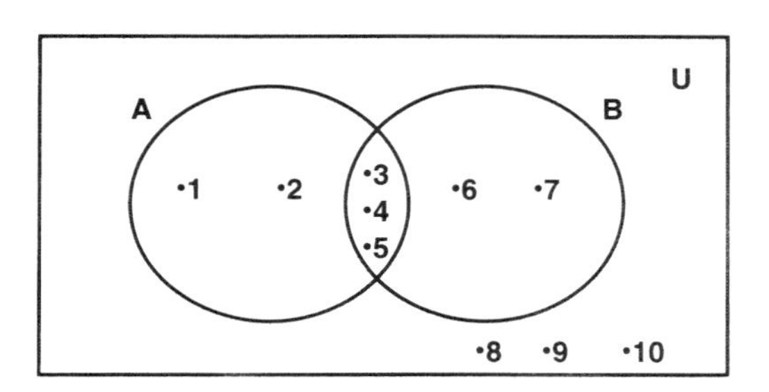

Exercise–1

3.

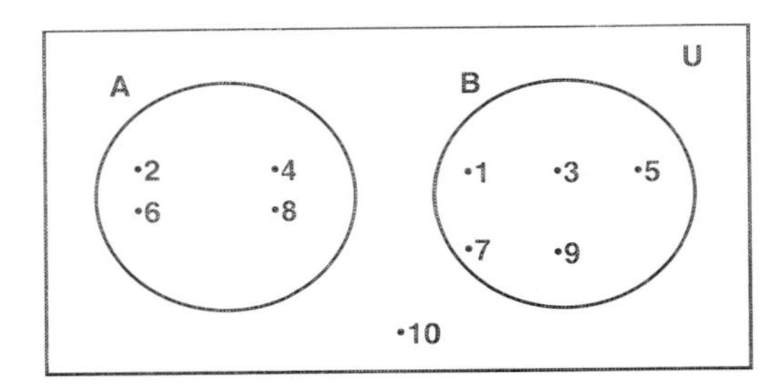

Exercise–3

5.

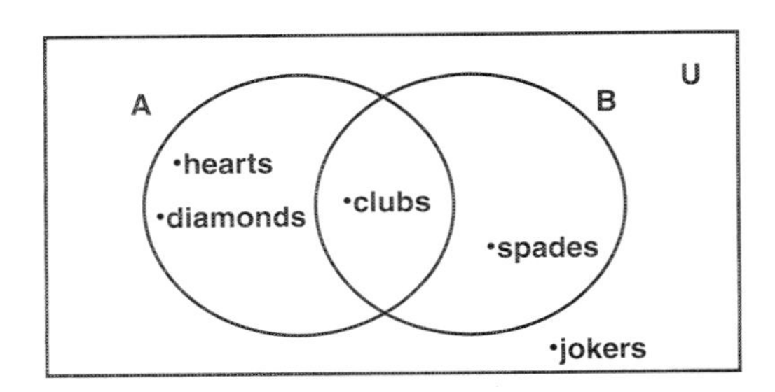

Exercise–5

7. $\{b,g,h,i,j,k\}$
9. $\{g,i\}$
11. $\{a,c,d,e,f\}$
13. $\{b,h,j,k,l,m\}$
15. $\{a,b,c,d,e,f,h,j,k,l,m\}$
17. $\{l,m\}$
19. $\{1,2,4,5,7\}$
21. $\{1,2\}$
23. $\{3,6\}$
25. $\{4,5,7,8,9\}$
27. $\{3,4,5,6,7,8,9\}$
29. $\{8,9\}$

EXERCISE SET #6

1a. 60 listeners
1b. 50 listeners
1c. 30 listeners
1d. 20 listeners
1e. 40 listeners

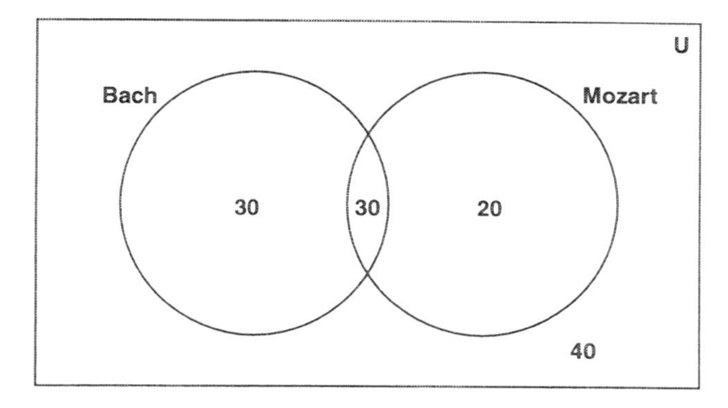

Exercise–1

3a. 620 students
3b. 320 students
3c. 390 students
3d. 90 students
3e. 290 students

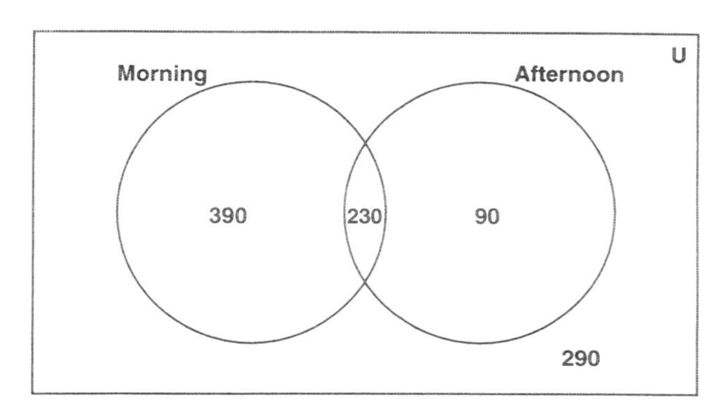

Exercise–3

5a. 220 patients
5b. 180 patients
5c. 50 patients
5d. 10 patients
5e. 20 patients

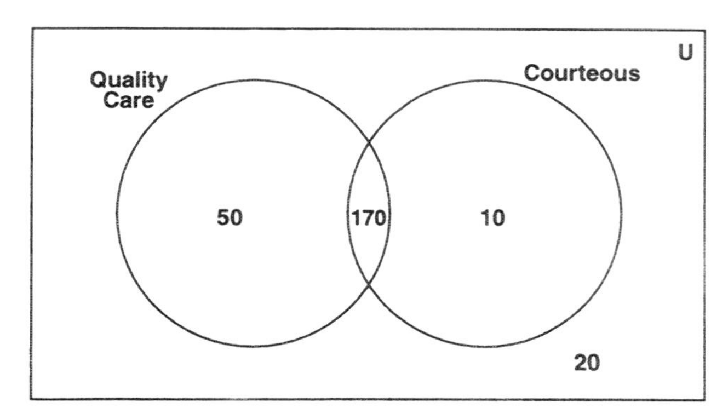

Exercise–5

7a. 340 customers 7b. 105 customers
7c. 211 customers 7d. 7 customers
7e. 115 customers 7f. 220 customers
7g. 320 customers 7h. 60 customers

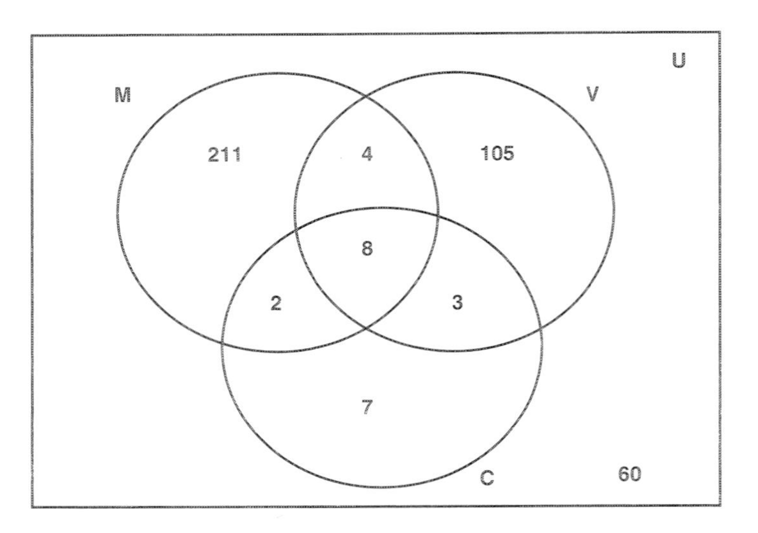

Exercise–7

EXERCISE SET # 7

1. $x = 12$
3. $x = 3$
5. $x = 2$
7. $x = 1$
9. $x = 5$
11. $x = -1$
13. $x = -24$
15. False
17. True
19. True
21. False
23. The solution of the equation $x = 2x$ is $x = 0$; thus, the student accidently divided by zero.
25. Assume to the contrary that both $ax_1 = b$ and $ax_2 = b$ hold true, where $x_1 \neq x_2$. Then we would have: $ax_1 = ax_2$ from which we get $x_1 = x_2$ – an obvious contradiction.
27. $x = 3$
29. $y = (a + b)/(a - b)$ provided that $a \neq b$.

EXERCISE SET # 8

1. $18 : 20 = \frac{18}{20} = \frac{9}{10}$
3. $540,000 : 500,000 = \frac{540000}{500000} = \frac{27}{25}$
5. $\$1200 : \$3600 = \frac{\$1200}{\$3600} = \frac{1}{3}$
7. $\$150,000 : \$35,000 = \frac{\$150000}{\$35000} = \frac{30}{7}$
9. $x = 28$
11. $x = 20$
13. $x = \frac{9}{2}$
15. 350 inches, or $29\frac{1}{6}$ feet.
19. height = 4
21. ($16/$10) · 10 customers = 16 customers.

EXERCISE SET # 9

1. #women = 14, #men = 19
3. 81 and 83
5. 4 nickels and 16 dimes.
7. 4 hrs. later, or at 4:00 p.m.
9. $16.00
11. 2 hrs later or at 3:00 p.m.
13. Speed of river's current = 5 miles per hour.
15. 200 shares of Stock-A, 250 shares of Stock-B.

EXERCISE SET # 10

1. $x < 2$
3. $x < -7$
5. $x \leq 2/5$
7. $x > 3$
9. $x > -1/2$
11. $x > 90$ where x is the score on the third quiz.
13. $x > 200$
15. We want $50x + 200 \leq 2400$, so that $x \leq 44$ boxes.
19. $-1 < x < 3$
21. $(a - b) < x < (a + b)$

EXERCISE SET # 11

1. Yes
3. Yes

5a. $g(0) = 0$

5b. $g(-1) = -1$

5c. $g(2) = 8$

5d. $g(1) = 1$

7. $Domain(f) = \{x \mid x \in R \text{ and } x \geq -1/2\}$
   $Range(f) = \{y \mid y \in R \text{ and } y \geq 0\}$
9. $Domain(f) = R$
   $Range(f) = \{y \mid y \in R \text{ and } y \geq -2\}$
11. $Domain(f) = R$
    $Range(f) = \{1\}$

13a. $C = \frac{5}{9}(F - 32)$ or $C = \frac{5}{9}F - \frac{160}{9}$

13b. $C = 100$ degrees Celsius

15a. $v = g(9.5) = 32(9.5) = 304$ feet per second

15b. $v = 207.27$ miles per hour.

17a. $D(30) = 100$

17b. It means that if corn were to be priced at $30 per bushel, then the demand would be 100 bushels.

19. If we assume that $x_1 < x_2$ then $ax_1 < ax_2$ since $a > 0$. Finally, adding b to both sides of the last inequality, we get $ax_1 + b < ax_2 + b$, or $f(x_1) < f(x_2)$ as required.
21. $A = f(w) = 400w$
23. $v = f(144) = \sqrt{64 \cdot 144} = 96$ feet per second, or 65.45 miles/hr.

EXERCISE SET # 12

5. $c = 5$
7. $c = \sqrt{5}$
9. $c = 2\sqrt{13} \approx 7.21$
11. $d = \sqrt{2}$
13. $d = 5$
15. $d = 5\sqrt{2}$
17. $\sqrt{12^2 + 16^2} = 20$ units from the origin.
21. midpoint: $(6, 5)$.
23. speed $= \left[\sqrt{(13-1)^2 + (18-2)^2}\right] \div 5 = 4$ feet per second.
25. Yes, since: $30^2 = 24^2 + 18^2 = 900$

EXERCISE SET # 13

1. True.
3. True.
5. False; its slope is 2..
7. True.
9. False; a vertical line cannot define a function.

11. $y = x + 2$

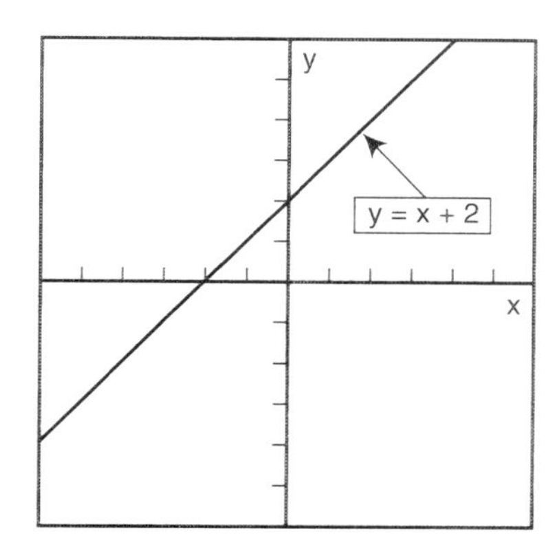

13. $y = \frac{1}{2}x + 1$

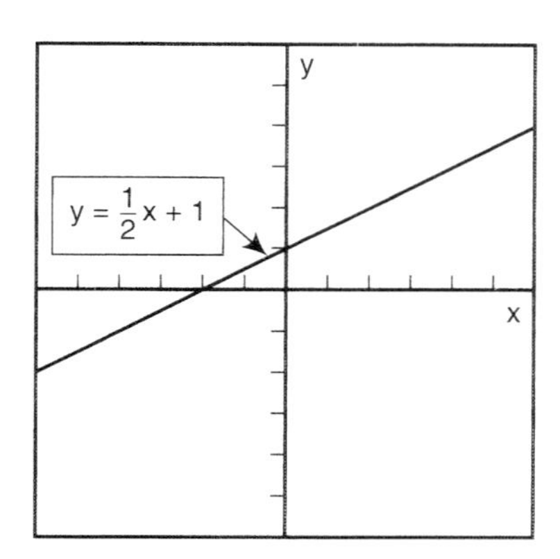

15. $y - 2x = 0$

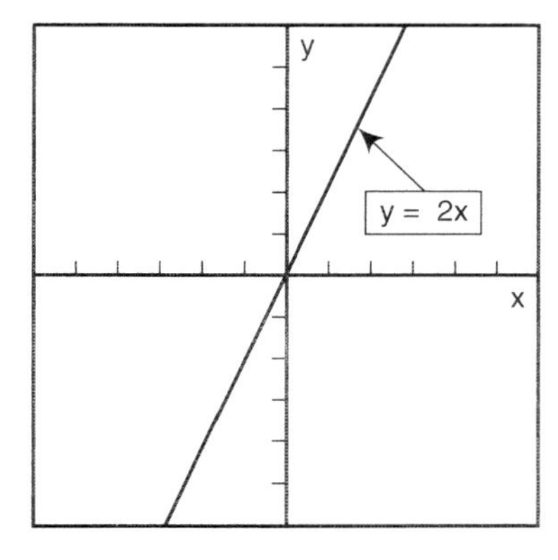

17. $y = -4x + 2$

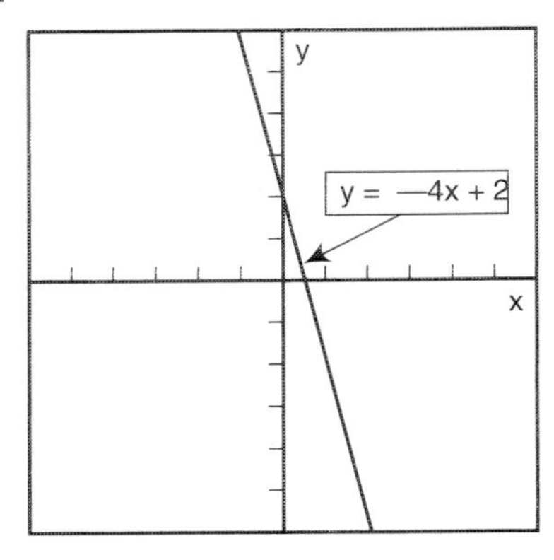

19. $m = 2/3$

21. m is undefined.

23. $m = 3/4$

25. $m = 3$

27. x-intercept (4,0), y-intercept (0,8)

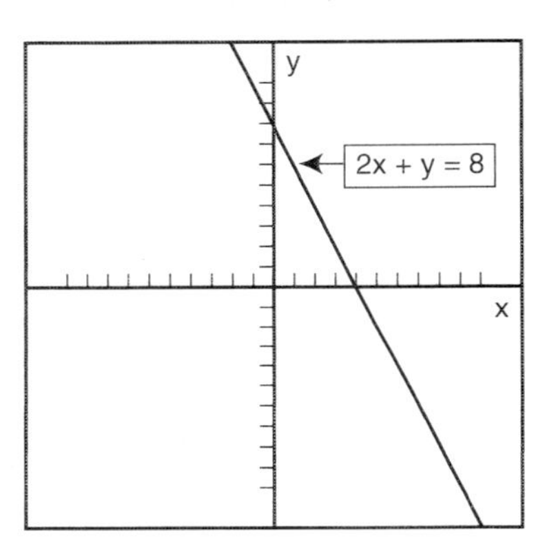

29. x-intercept (2,0), y-intercept (0,6)

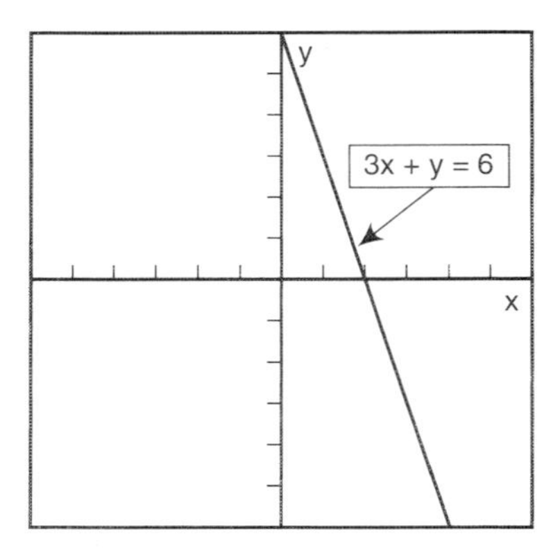

31. x-intercept (6,0), y-intercept (0,3)

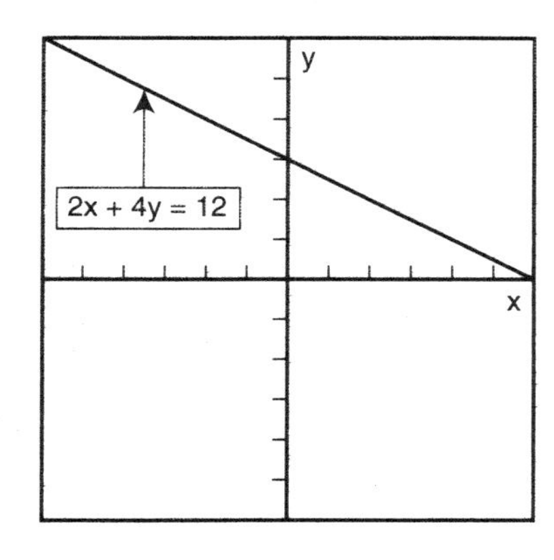

33. $m = 2$, y-intercept: (0,2)
35. $m = 2$, y-intercept: (0,−2)
37. $m = 0$, y-intercept: (0,3)
39. $m = 2$, y-intercept: (0,−2)
41. $y = -2x + 3$
45. $y = -2x + 9$

EXERCISE SET # 14

1. $x = 1, y = 1$
3. Inconsistent system (no solutions)
5. $x = 1, y = 2$
7. $x = 0, y = 0$
9. Inconsistent system (no solutions)
11. True
13. True
15. False
17a. $2x + 2y = 12, x - 2y = 0$
17b. $x = 4, y = 2$ [length = 4 miles, width = 2 miles]
19a. $x + y = 5, 9x - 9y = -9$
19b. $x = 2, y = 3$ so the number is 23.
21a. $x + y = 23, x - y = 7$
21b. $x = 15$ miles (running), $y = 8$ miles (swimming)
23a. $x + y = 33, 2x - y = -3$
23b. $x = 10, y = 23$
25. Shirt costs $10, trousers cost $32

EXERCISE SET # 15

1. $x = -1, y = 1$
3. $x = 2, y = 3$
5. $x = 5, y = 4$
7. $x = -2, y = -2$

9. $x = 3/20, y = -13/15$
11. $x = -1, y = 1$
13. $x = 1/3, y = -1/3$
15. $x = 11/9, y = -13/18$
17. $x = 0, y = -1$
19. $x = 17/11, y = -30/11$
21. $x = 5$ nickels, $y = 10$ dimes
23. We obtain the system: $x+y=7,\ x+10y = \frac{1}{2}(10x+y)-1$ whose solution is $x = 5, y = 2$. Thus, the original number is 52.
25. Cassette tapes cost $8 each, CD's cost $10 each.

EXERCISE SET #, 16

1a. $P(\text{drunk}) = 5400/11723 = .461$
1b. $P(\text{not drunk}) = (11723 - 5400)/11723 = .539$
3a. $P(\text{hit}) = 3283/10881 = .302$
3b. $P(\text{homerun}) = 660/10881 = .061$
5a. $P(\text{excellent}) = 104/200 = .52$
5b. $P(\text{good}) = 60/200 = .30$
5c. $P(\text{no opinion}) = 6/200 = .03$
7a. $P(\text{win}) = 18/120 = .15$
7b. $P(\text{lose}) = (120 - 18)/120 = .85$
9a. $P(\text{fair}) = 120/500 = .24$
9b. $P(\text{good}) = 255/500 = .51$
9c. $P(\text{poor}) = 75/500 = .15$

EXERCISE SET #, 17

1. $P(\text{queen}) = 4/52 = \frac{1}{13}$
3. $P(\text{ten}) = 4/52 = \frac{1}{13}$
5. $P(\text{black card}) = 26/52 = \frac{1}{2}$
7. $P(\text{two}) = 4/52 = \frac{1}{13}$
9. $P(\text{a card greater than 2 but less than 5}) = 8/52 = \frac{2}{13}$
11. $P(\text{not extremely satisfied}) = 1 - 0.95 = .05$
13a. $P(\text{hotel major}) = 10/30 = \frac{1}{3}$

13b. P(culinary major) = 5/30 = $\frac{1}{6}$
13c. P(business major) = 15/30 = $\frac{1}{2}$
13d. P(not a hotel major) = 1 − 1/3 = $\frac{2}{3}$
13e. P(not a business major) = 1 − 1/2 = $\frac{1}{2}$
15a. P(2 heads) = 1/4
15b. P(0 heads) = 1/4
15c. P(1 head) = 1/2
15d. P(at least one head) = 3/4

EXERCISE SET # 18

1. P(even or odd number) = 1
3. P(even number or number < 4) = $\frac{5}{6}$
5. P(2 or 5) = $\frac{1}{3}$
7. P(odd number or a 4) = $\frac{2}{3}$
9. P(number greater > 3 or < 5) = 1
11. P(jack or 3) = $\frac{2}{13}$
13. P(queen or red card) = $\frac{7}{13}$
15. P(heart or diamonds) = $\frac{1}{2}$
17. P(jack or club) = $\frac{4}{13}$
19. P(10 or a jack) = $\frac{2}{13}$

21a. P(orange or apple) = $\frac{3}{4}$
21b. P(apple or peach) = $\frac{5}{6}$
21c. P(orange or peach) = $\frac{5}{12}$
23a. P(likes service) = $\frac{40}{100} = \frac{2}{5}$
23b. P(likes room) = $\frac{50}{100} = \frac{1}{2}$
23c. P(likes both service and room) = $\frac{20}{100} = \frac{1}{5}$
23d. P(likes room or service) = $\frac{7}{10}$

EXERCISE SET # 19

1. $P(4 \mid \text{even}) = \frac{1}{3}$
3. $P(3 \mid \text{odd}) = \frac{1}{3}$
5. $P(3 \mid \text{number} < 5) = \frac{1}{4}$
7. $P(6 \mid \text{number} > 4) = \frac{1}{2}$
9. $P(6 \mid \text{number is divisible by } 3) = \frac{1}{2}$
11. $P(\text{diamond} \mid \text{red card}) = \frac{1}{2}$
13. $P(2 \text{ of diamonds} \mid \text{red card}) = \frac{1}{26}$
15. $P(\text{ace of clubs} \mid \text{card is a club}) = \frac{1}{13}$
17. $P(\text{face card} \mid \text{red card}) = \frac{3}{13}$
19. $P(\text{queen} \mid \text{face card}) = \frac{1}{3}$

21b. $P(\text{2nd child is boy} \mid \text{1st child is girl}) = \frac{1}{2}$

21c. $P(\text{2nd child is boy} \mid \text{1st child is boy}) = \frac{1}{2}$

21d. $P(\text{2nd child is girl} \mid \text{1st child is boy}) = \frac{1}{2}$

23. $P(\text{add up to } 7 \mid \text{first die is } 5) = \frac{1}{6}$
25. $P(\text{add up to } 6 \mid \text{first die is } 3) = \frac{1}{6}$
27. $P(\text{sum} < 4 \mid \text{first die is } 2) = \frac{1}{6}$

29a. $P(\text{accomodations good} \mid \text{non - business guest}) = \frac{5}{8}$

29b. $P(\text{accomodations poor} \mid \text{business guest}) = \frac{1}{8}$

29c. $P(\text{guest feels accomodations are good}) = \frac{3}{4}$

29d. $P(\text{business guest} \mid \text{accomodations poor}) = \frac{1}{4}$

31a. $P(2 \text{ both times}) = \frac{1}{36}$

31b. $P(\text{even number both times}) = \frac{1}{4}$

31c. $P(2 \text{ odd number}) = \frac{1}{4}$

33a. $P(2 \text{ hits}) = .9025$

33b. $P(2 \text{ misses}) = .0025$

33c. $P(\text{at least one hit}) = .9975$

37. $P(\#\text{heads} \geq 1) = 1 - P(0 \text{ heads in } n \text{ tosses}) = 1 - \left(\frac{1}{2}\right)^n$

    As n increases, this probability approaches 1, as it should.

EXERCISE SET # 20

1. $E(X) = 5.5$
3. $E(X) = 25$

5a. $E(\text{winnings}) = -\$\frac{7}{4} = -\$1.75$

5b. $E(\text{church earnings}) = \$800 - \$100 = \$700$

5c. Fair price is $p = \$\frac{1}{4} = \$00.25$

7a. $E(\text{winnings}) = -\$\frac{15}{38} \approx -\$00.395$

7b. Fair price is $p = \$\frac{175}{38} \approx \$4.61$

7c. $3.95

9a. $E(\text{winnings}) \approx -\$00.62$

9b. Fair price is $p \approx \$00.38$

11. $E(X) = \frac{1}{N}(1) + \frac{1}{N}(2) + \ldots + \frac{1}{N}(N) = \frac{1}{N}(1 + 2 + \ldots + N)$

$= \frac{1}{N}\frac{N(N+1)}{2} = \frac{N+1}{2}$ as required.

13. $E(X) = 3.45$
15. $E(X) = 10$

EXERCISE SET # 21

1. $I = \$516$
3. $I = \$36$
5. $I = \$36.83$
7. $I = \$903.38$
9. $I = \$806.40$
11. $A = \$13{,}203.13$
13. $r = 19.23\%$

15a. $D = \$1248$

15b. $P = \$3952$

17. $P = \$1705.50$
19. $P = \$5{,}760$

EXERCISE SET # 22

1. $A = \$1405.99$
3. $A = \$2090.65$
5. $A = \$7641.93$
7. $P = \$2717.35$
9. $P = \$4617.42$
11. $A = \$14038.30$
13. $r_E = .0824$

15a. r_E(Happy) $= .06136$
15b. r_E(Peoples) $= .06296$
15c. Peoples Credit Union

17. $A \approx \$1336$
19. $i = 18.92\%$

EXERCISE SET # 23

1a. $A = \$2253.06$
1b. Interest $= \$253.06$

3. $A = \$611.12$
5. $A = \$259{,}056$
7. $R = \$50.69$
9. $P = \$33550.41$
11. $R = \$95.69$
15. $R = \$486.63$

1. 45%
3. 112%
5. 6%
7. 0.25%
9. 0.42
11. 0.002
13. 0.00001
15. 0.734
17. 150
19. 18
21. 105
23. 5%
25. 160%
27. 250
29. $333^{1/3}$

31. {x | x is both a customer of IBM and a member of the board of directors of IBM}

33. $p(15, 4) = 32{,}760$

35.

37. (a) 40 (b) 35 (c) 40 (d) 205 (e) 155

38. 46

39. \$2, \$3.50 \$15.50 \$36.50 \$2 for the first hour and \$1.50 for each additional hour

43. (a) \$25 (b) \$15.00 (c) Goes to 0

(d)

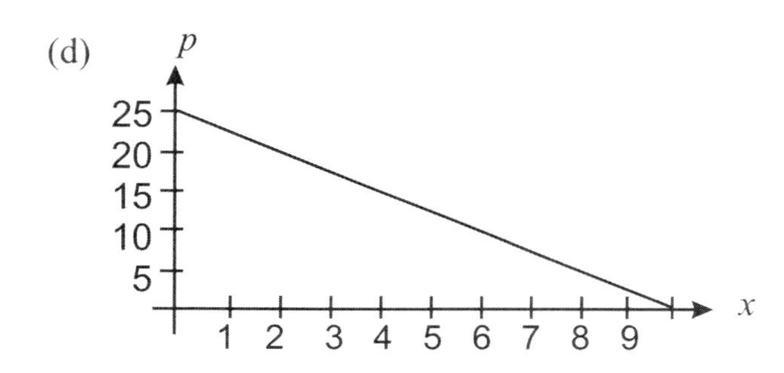

44. (a) $p = 0.8x - 50$ (b) 63 (c) 110

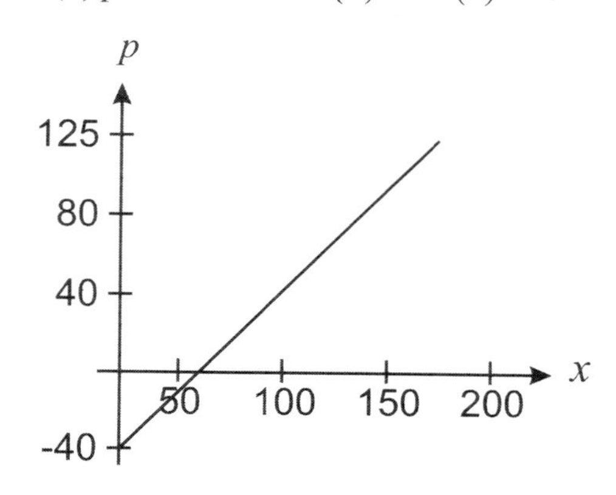

45. (a) $\frac{25}{2}$ (b) $\frac{15}{2}$

(c) 0; it would take a negative price to generate a demand >10.

(d)

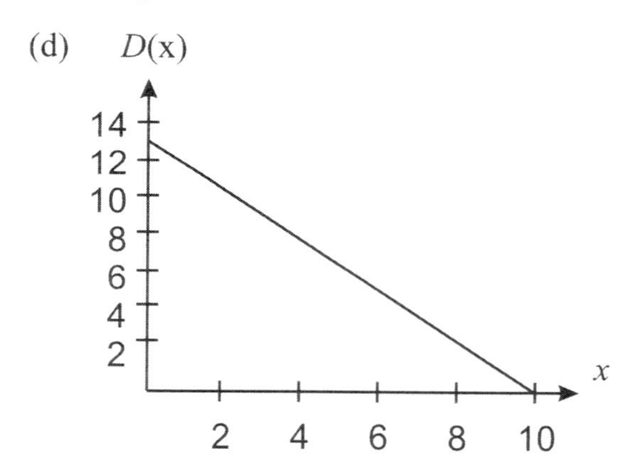

47. (a)

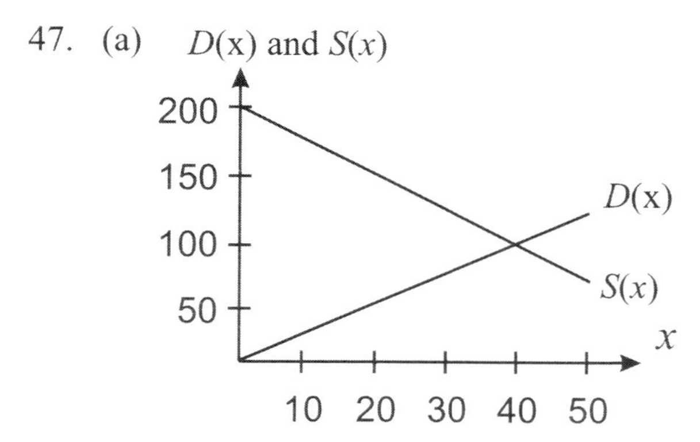

(b) (40, 100)
(c) $p = 100$
(d) $x > 40$

49. 1400 items at a cost of \$350

51. Yes. The break-even point will be reached at a lower level of production

52. $C = 0.2x + 280$ dollars

54. 1200 items

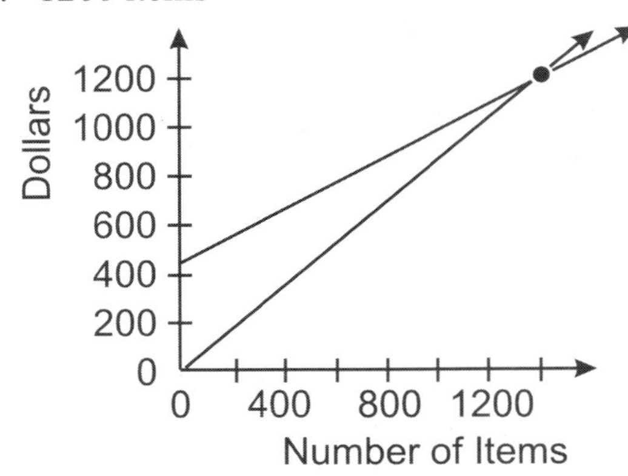

56. 200 newspapers

60. (a) .90 (b) .1 (c) $.97

62. (a) $\frac{44}{120} = \frac{11}{30}$ (b) $\frac{7}{120}$

(c) $\frac{76}{120} = \frac{19}{30}$ (d) $\frac{23}{120}$

65. $\frac{1}{2}$

67. (a) .75750 (b) .7225

70. (a) $187.50 (b) $40 (c) $2933.33

72. (a) $80 (b) $36 (c) $100

74. $I = \$360$, $A = \$3360$

76. (a) $I = \$720.98$, $A = \$2720.98$
(b) $I = \$737.14$, $A = \$2737.14$
(c) $I = \$745.57$, $A = \$2745.57$

79. (a) $A = \$4480$ (b) $A = \$3075$ (c) $A = \$124$

80. $3736.29

82. $5525.86

84. Between 18.25 and 18.5 years; ≈18.45 years

86. $133.33

89. 8.16%

91. 8.3%

93. $A = \$14{,}917.92$

95. $5866.60

98. $273.02

99. $10,810.76

101. 9.07%

103. Approx. $15^{1/4}$ yr

106. (a) $15,200 (b) $60,800 (c) $498.21
(d) $115,315.60 (e) $199 mo = 16yr 7 mo
(f) $56,452.79

107. (a) 545.22 (b) $130,566 (c) $23,501.79

108. Monthly payment: $1049; Equity after 10 yr: $21,575.51

INDEX

G

H

I

L

M

N

O

P

R

S

T

U

V

X

Y